The Molecular Biology of Hormone Action

THE MOLECULAR BIOLOGY OF HORMONE ACTION
The Thirty-Fourth Symposium of
The Society for Developmental Biology

Orono, Maine, June 1-4, 1975

EXECUTIVE COMMITTEE

1974-1975

DONALD D. BROWN, Carnegie Institution of Washington, *President*
ELIZABETH D. HAY, Harvard Medical School, *Past-President*
WILLIAM J. RUTTER, University of California, *President-Designate*
JAMES A. WESTON, University of Oregon, *Secretary*
JOHN PAPACONSTANTINOU, Oak Ridge National Laboratory, *Treasurer*
VIRGINIA WALBOT, Washington University, *Member-at-Large*

1975-1976

WILLIAM J. RUTTER, University of California, *President*
DONALD D. BROWN, Carnegie Institution of Washington, *Past-President*
IAN M. SUSSEX, Yale University, *President-Designate*
JAMES A. WESTON, University of Oregon, *Secretary*
MARIE DIBERARDINO, Medical College of Pennsylvania, *Treasurer*
VIRGINIA WALBOT, Washington University, *Member-at-Large*

Business Manager
CLAUDIA FORET
P.O. BOX 43
Eliot, Maine 03903

The Molecular Biology of Hormone Action

Edited by

John Papaconstantinou

*Biology Division
Oak Ridge National Laboratory
Oak Ridge, Tennessee*

1976

ACADEMIC PRESS, INC.   New York   San Francisco   London
A Subsidiary of Harcourt Brace Jovanovich, Publishers

COPYRIGHT © 1976, BY ACADEMIC PRESS, INC.
ALL RIGHTS RESERVED.
NO PART OF THIS PUBLICATION MAY BE REPRODUCED OR
TRANSMITTED IN ANY FORM OR BY ANY MEANS, ELECTRONIC
OR MECHANICAL, INCLUDING PHOTOCOPY, RECORDING, OR ANY
INFORMATION STORAGE AND RETRIEVAL SYSTEM, WITHOUT
PERMISSION IN WRITING FROM THE PUBLISHER.

ACADEMIC PRESS, INC.
111 Fifth Avenue, New York, New York 10003

United Kingdom Edition published by
ACADEMIC PRESS, INC. (LONDON) LTD.
24/28 Oval Road, London NW1

Library of Congress Cataloging in Publication Data

Society for Developmental Biology.
    The molecular biology of hormone action.

    (Symposium of the Society for Developmental Biology ; 34th)
    Papers presented at the Society's 34th symposium held in Orono, Me., June 1-4, 1975.
    1. Steroid hormones–Congresses. 2. Genetic transcription–Congresses. 3. Eukaryotic cells–Congresses.
4. Hormones (Plants)–Congresses. I. Papaconstantinou, John. II. Title. III. Series: Society for Developmental Biology. Symposium ; 34th. [DNLM: 1. Molecular biology–Congresses. 2. Steroids–Congresses. W1 S0851M no. 34 1975 / WK150 S679m 1975]
QH511.S6 no. 34   [QP572.S7]   574.3'08s   [574.1'927]
ISBN 0-12-612980-0   76-47450

PRINTED IN THE UNITED STATES OF AMERICA

Contents

LIST OF CONTRIBUTORS vii
FOREWORD... xi
ACKNOWLEDGMENTS.. xiii

I. Transcription in Eukaryotic Cells

Organization of Proteins in Chromatin 3
   G. Felsenfeld, B. Sollner-Webb, D. Camerini-Otero, and W. Melchior

The Role of Non-Histone Proteins in the Regulation of Globin Gene Activity 15
   R. Stewart Gilmour and J. MacGillivray

Function, Structure, and Regulation of Eukaryotic Nuclear RNA Polymerases 29
   Robert G. Roeder, Lawrence B. Schwartz, and V. E. F. Sklar

Biochemical Events in mRNA Formation in Mammalian Cells ... 53
   James E. Darnell, Warren Jelinek, Larry Puckett, Eva Derman, and Steven Bachenheimer

Synthesis and Translation of Messenger RNA during Differentiation of the Cellular Slime Mold *Dictyostelium Discoideum* 75
   Harvey F. Lodish, Tom Alton, Robert P. Dottin, Alan M. Weiner, and Jeanne P. Margolskee

II. Action of Steroid Hormones

Regulation of Transcription of the Eucaryotic Genome 107
   H. C. Towle, M.-J. Tsai, M. Hirose, S. Y. Tsai, R. J. Schwartz, M. G. Parker, and B. W. O'Malley

Estrogen-Receptor Movement to the Nucleus: Discussion of a Cytoplasmic-Exclusion Hypothesis......................... 137
   Frank Gannon, Benita Katzenellenbogen, George Stancel, and Jack Gorski

Investigation of the Mechanisms and Consequences of Steroid
Hormone Action on Vitellogenin Synthesis in *Xenopus laevis* .. 151
  Lawrence J. Wangh, Ruth F. Longthrone, and John Knowland

III. Action of Plant Hormones

The Role of Hormones in the Integration of Seedling Growth ... 173
  J.E. Varner and David Tuan-Hua Ho

Subject Index 195

List of Contributors

Boldface Denotes Chairmen

Tom Alton, Department of Biology, Massachusetts Institute of Technology, Cambridge, Massachusetts

Steven Bachenheimer, Rockefeller University, New York, New York

D. Camerini-Otero, Laboratory of Molecular Biology, National Institute of Arthritis, Metabolism and Digestive Diseases, National Institutes of Health, Bethesda, Maryland

James E. Darnell, Rockefeller University, New York, New York

Eva Derman, Rockefeller University, New York, New York

Robert P. Dottin, Department of Biology, Massachusetts Institute of Technology, Cambridge, Massachusetts

Sarah C. R. Elgin, Biological Laboratories, Harvard University, Cambridge, Massachusetts

G. Felsenfeld, Laboratory for Molecular Biology, National Institute of Arthritis, Metabolism and Digestive Diseases, National Institutes of Health, Bethesda, Maryland

Frank Gannon, Department of Biochemistry, University of Wisconsin, Madison, Wisconsin

R. Stewart Gilmour, Beatson Institute for Cancer Research, 132 Hill Street, Glasgow, Scotland

Jack Gorski, Department of Biochemistry, University of Wisconsin, Madison, Wisconsin

M. Hirose, Department of Cell Biology, Baylor College of Medicine, Houston, Texas

David Tuan-Hua Ho, MSU/AEC Plant Research Laboratory, Michigan State University, East Lansing, Michigan

Warren Jelinek, Rockefeller University, New York, New York

LIST OF CONTRIBUTORS

Elwood V. Jensen, Ben May Laboratory for Cancer Research, The University of Chicago, Chicago, Illinois

Benita Katzenellenbogen, Departments of Physiology and Biophysics, University of Illinois, Urbana, Illinois

Joe L. Key, Biological Sciences, University of Georgia, Athens, Georgia

John Knowland, Department of Anatomy, The Medical School, University Walk, Bristol, England

Harvey F. Lodish, Department of Biology, Massachusetts Institute of Technology, Cambridge, Massachusetts

Ruth F. Longthorne, MRC Laboratory of Molecular Biology, Cambridge, England

A. J. MacGillivray, Beatson Institute of Cancer Research, 132 Hill Street, Glasgow, Scotland

Jeanne P. Margolskee, Department of Biology, Massachusetts Institute of Technology, Cambridge, Massachusetts

W. Melchior, Laboratory of Molecular Biology, National Institute of Arthritis, Metabolism and Digestive Diseases, National Institutes of Health, Bethesda, Maryland

B. W. O'Malley, Department of Cell Biology, Baylor College of Medicine, Houston, Texas

M. G. Parker, Department of Cell Biology, Baylor College of Medicine, Houston, Texas

Larry Puckett, Rockefeller University, New York, New York

Lynn M. Riddiford, University of Washington, Seattle, Washington

Robert G. Roeder, Department of Biochemistry, Washington University School of Medicine, St. Louis, Missouri

William J. Rutter, Department of Biochemistry, University of California, San Francisco Medical Center, San Francisco, California

Lawrence B. Schwartz, Department of Biological Chemistry, Washington University, St. Louis, Missouri

R. J. Schwartz, Department of Cell Biology, Baylor College of Medicine, Houston, Texas

LIST OF CONTRIBUTORS

V. E. F. Sklar, Department of Biological Chemistry, Washington University, St. Louis, Missouri

B. Sollner-Webb, Laboratory for Molecular Biology, National Institute of Arthritis, Metabolism and Digestive Diseases, National Institutes of Health, Bethesda, Maryland

George Stancel, Program in Pharmacology, University of Texas Medical School at Houston, Texas Medical Center, Houston, Texas

H. C. Towle, Department of Cell Biology, Baylor College of Medicine, Houston, Texas

M.-J. Tsai, Department of Cell Biology, Baylor College of Medicine, Houston, Texas

S. Y. Tsai, Department of Cell Biology, Baylor College of Medicine, Houston, Texas

J. E. Varner, Department of Biology, Washington University, St. Louis, Missouri

Lawrence J. Wangh, National Institute for Medical Research, Mill Hill, London, England

Alan M. Weiner, Department of Biology, Massachusetts Institute of Technology, Cambridge, Massachusetts

Foreword

The 34th Annual Symposium of the Society brought together experts studying the mechanisms of hormone action in plants and animals with those analyzing gene action in eukaryotes. The topics and speakers were chosen by Professor Elwood Jensen of the University of Chicago with an assist from the Board.

Developmental biologists who search for embryonic "determinants" and trace growth factors may be looking for molecules that act by mechanisms that are analogous to the protein and steroid hormones with which we are all familiar. There is a vast literature on the physiology and pharmacology of many hormones, including an array of their specific analogs and inhibitors. We take for granted the fact that important hormones, with known developmental effects on tissues, have been purified, characterized chemically, and are available in pure form from the chemical supply houses. So while the search for determinants continues, hormone research has long since advanced to an analysis of mechanism. Indeed, elucidation of the first step of hormone action, their interaction with cell receptors, constitutes a major scientific breakthrough in modern biology. The Society's Board chose this burgeoning field in recognition of its relevance to cell specialization. Since current research suggests that hormone-receptor complexes directly or indirectly affect gene function we devoted a session of the Symposium to our current knowledge of the control of eukaryotic gene expression. Perhaps we hastened the already rapid convergence of these two related areas.

We broke with Society tradition by not following the usual "molecule to organism" format. Molecules predominated overwhelmingly, and we encouraged the speakers to discuss the latest findings, in contrast with general reviews. The Board felt that since no more than 10 to 20% of the members attend any one Symposium, we might appeal to a different segment of the membership with this approach. In my opinion, variations on this and other themes should be experimented with in future annual Symposia of the Society.

Several hundred enthusiastic participants enjoyed good science mingled with the superb hospitality of the University of Maine. All local details were executed flawlessly by Dr. Alton Mun, and his committee from the Zoology Department.

The proceedings were marred by the shocking news of Gordon Tomkins' illness, which ended in his untimely death. He was to have presented the recent results of his laboratory, a group that pioneered many advances in hormone research. Gordon was a man of intellectual breadth and humanistic values—a unique individual. We wish to acknowledge his major contributions to the subject of this Symposium.

Donald D. Brown

Acknowledgments

The Symposium was funded in part by generous grants from the National Cancer Institute and the National Science Foundation. We received additional support from Upjohn Company, Merck, Sharpe and Dohme, Plenum Press, Oxford University Press, Spectra Instruments, Atlantex and Ziela Instruments, Wiley, Alexander Scientific Corporation, and a Sea Life Grant to the University of Maine.

# I. Transcription in Eukaryotic Cells

ORGANIZATION OF PROTEINS IN CHROMATIN

G. Felsenfeld, B. Sollner-Webb
D. Camerini-Otero and W. Melchior

Laboratory of Molecular Biology
National Institute of Arthritis,
Metabolism and Digestive Diseases
National Institutes of Health
Bethesda, Maryland 20014

I. INTRODUCTION

Recent electron microscopic studies of chromatin (Olins and Olins, 1974; Oudet *et al.*, 1975) reveal the presence of repeating structural units in the form of "beads on a string". The nucleoprotein beads have a diameter of about 100 Å, and are connected by strands of DNA which may be thinly covered by protein. Further evidence for a subunit structure comes from studies of the nuclease digestion of nuclei and chromatin. Rill and Van Holde (1973) found that mild nuclease treatment of chromatin gave rise to a homogeneous population of small nucleoprotein fragments. Hewish and Burgoyne (1973) showed that autodigestion of nuclei by an endogenous nuclease led to the release of DNA fragments that were multiples of a unit 200 base pairs long. Noll (1974) treated nuclei with staphylococcal nuclease and obtained similar fragments. On the basis of these results, Kornberg (1974) has proposed that chromatin is predominantly organized in repeating units, involving about 200 base pairs of DNA and each containing an identical complement of histone molecules: two each of the arginine-rich pair (H3 and H4) and the slightly lysine-rich pair (H2A and H2B).

In discussing the "structure" of chromatin, it is useful to make a distinction between the gross structure characteristic of the bulk of the material, and the structure in the neighborhood of genes which are being actively transcribed. Considerable effort has been expended in trying to demonstrate a correlation between template activity and structure. It has been suggested by a number of investigators that the actively transcribed chromatin has a relatively more extended structure than inactive chromatin, though no clear cut relationship of this kind has been established for an identifiable gene known to be active *in vivo*. The relationship between the overall structure of chromatin and the arrangement of proteins in the neighborhood of actively transcribed genes remains unresolved.

In this paper we describe studies of gross chromatin structure which employ chemical probes; we then apply similar techniques to an examination of protein distribution in the neighborhood of an actively transcribed gene.

II. GROSS STRUCTURE OF CHROMATIN

It was shown some years ago in our laboratory (Clark & Felsenfeld, 1971) that when chromatin isolated from calf thymus or other tissues is digested with staphylococcal nuclease, the reaction proceeds to a limit in which about half the DNA is digested. The DNA protected from digestion consists of small double stranded fragments with a weight average size of 110 base pairs (Clark & Felsenfeld, 1974). What is the relationship of this limit digest pattern to the fragments produced by mild digestion of nuclei with staphylococcal nuclease? To determine this we have examined the entire process of digestion of nuclei, from early times to the limit (Sollner-Webb & Felsenfeld, 1975). The kinetics of digestion of nuclear DNA by staphylococcal nuclease are shown in Fig. 1. It should be noted that the kinetics are identical to those obtained when chromatin is used as a substrate. If the reaction is stopped when a few per

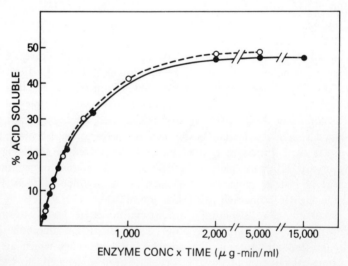

Fig. 1. *Kinetics of nuclear digestion-release of acid soluble products. Duck reticulocyte nuclei (•) and duck reticulocyte chromatin (o) were digested with 1 to 60 μg/ml staphylococcal nuclease for 10 min to 1 hr in 1 mM Tris pH 8, 0.1 mM $CaCl_2$ at 1 mg DNA/ml. The fraction of the DNA soluble in 0.4 M perchloric acid, 0.4 M NaCl was measured. RNA was shown to contribute <3% of the A_{260}. (From Sollner-Webb & Felsenfeld, 1975).*

cent of the nuclear DNA is acid soluble, and the undigested DNA is extracted and analyzed by gel electrophoresis, a series of discrete bands are detected, which are multiples of a fundamental subunit 185 base pairs in length (Fig. 2). This result is essentially the same one reported by Hewish & Burgoyne (1973) and by Noll (1974). In a manner similar to that reported by Noll, we are able to separate nucleoprotein fractions by sedimentation on sucrose gradients (Fig. 3) and show that more rapidly sedimenting fractions contain the higher multiples of the fundamental DNA subunit (Fig. 4). The fractions thus correspond to monomer, dimer, trimer and higher order nucleoprotein structures. If the purified trimer fraction is again treated with nuclease, it is degraded to dimer and monomer. Purified dimer, similarly treated, is reduced to monomer (Sollner-Webb & Felsenfeld, 1975). The monomer is thus an intermediate form through which most of the chromatin must pass in the course of digestion. It presumably corresponds to the "bead and string" discussed in the Introduction.

The monomer and its associated 185 base-pair DNA (referred to as band I in our gels) is not indefinitely stable, but is subject to further digestion. In the next step, band I decreases slightly in size to form a sharp band, IA, with a

Fig. 2. *Polyacrylamide gel electrophoresis of nuclear digest DNA. Nuclei were digested and the isolated DNA was run on 3% disc gels. From left to right are 1, 2, 3, 8, 16, 25, and 48% acid soluble digest. The last gel is a chromatin limit, 48%, digest. (From Sollner-Webb & Felsenfeld, 1975).*

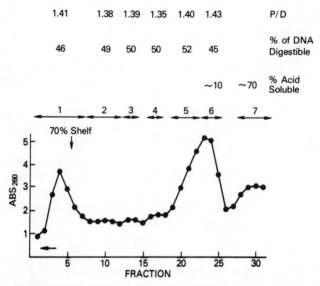

Fig. 3. *Sucrose gradient centrifugation of partially digested nuclei. Nuclei were partially digested (7% acid soluble DNA), the reaction was stopped by the addition of Na-EDTA to 0.3 mM, and the mixture was layered onto sucrose gradients. Direction of sedimentation is indicated by an arrow. The median sedimentation coefficient of the peak monomer fractions (S_{20}) was 10.2, measured in 1 mM Tris-HCl, 0.1 mM Na-EDTA, pH 8 in the Model E ultra-centrifuge. (From Sollner-Webb & Felsenfeld, 1975).*

size of 140 base pairs. We believe that this transition from I to IA involves the removal of relatively accessible DNA "strings" about 45 base pairs in length leaving a fairly homogeneous "bead" containing 140 base pairs of DNA.

This bead is itself susceptible to further digestion, ultimately giving rise to a series of discrete double-stranded DNA fragments ranging in size from about 140 to 45 base pairs which comprise the nuclear limit digest. The complete sequence of events in nuclear digestion is summarized in Fig. 5. When similar experiments are carried out with chromatin purified from nuclei, the oligomeric nucleoprotein intermediates are observed, but the corresponding DNA oligomeric intermediates may or may not be present, depending upon the method of chromatin preparation (Sollner-Webb & Felsenfeld, 1975). Certain methods of preparing chromatin apparently allow the nuclease to nick the DNA within the beads at a rate comparable to the rate of nicking of the strings; although the protein continues to hold the bead structure together, its removal allows one to observe the fragmentation of the DNA in the repeating structure. Nonetheless, digestion of chromatin to the limit yields a digest pattern quite similar to that observed with nuclei (Fig. 6), except for the presence of fragments of sizes as large as 157 base pairs in length.

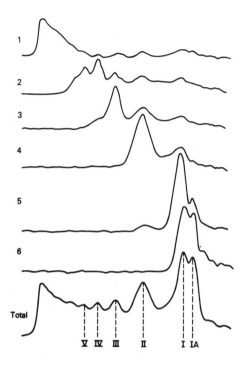

Fig. 4. *Polyacrylamide gels of DNA from isolated nucleoprotein fragments. DNA isolated from the sucrose gradient fractions, pooled as indicated in Fig. 4, was run on 3% acrylamide gels, stained, and photographed. Negatives were scanned with a Joyce-Loebl microdensitometer. These scans have been shown to be linear in DNA concentration. Migration is from left to right. Scan numbers indicate the pooled fractions; "Total" is DNA isolated from the material that was layered onto the sucrose gradient. (From Sollner-Webb & Felsenfeld, 1975).*

What is the origin of these fragments? We have carried out experiments (Axel et al., 1974) in which duck reticulocyte histones are completely dissociated from their DNA and then reconstituted either with duck DNA or with a bacterial DNA. In either case a limit digest of the reconstitute yields a band pattern similar to that obtained with native chromatin. We conclude that the pattern arises from specific contacts between histones and DNA, which are determined by the histone structure rather than the DNA base composition. To demonstrate further the role of the histones in band formation, we have reconstituted duck DNA in the presence of individual purified duck histones and combinations of histones. Under our conditions of reconstitution, a single histone, although it protects DNA segments from digestion, does not give rise to a discrete band pattern, but rather to a broad continuous distribution of segment sizes. However, the arginine rich histone pair, H3 and H4, gives rise to the lower molecular weight DNA bands in such reconstitution experiments, and if the slightly lysine-rich pair H2A and H2B is also added to the reconstitution mixture, most of the higher molecular weight bands (157-80 base pairs) of the chromatin limit digest pattern also appear (Fig. 7). On the other hand, if slightly lysine-rich histones alone are added to DNA, the limit digest contains no discrete bands (Fig. 8). The higher molecular bands only appear if arginine-rich histone is present as well. Our most recent results

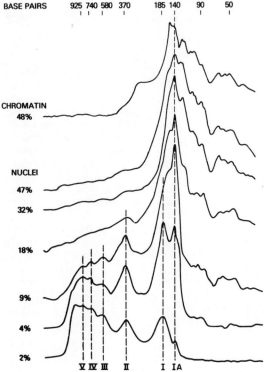

Fig. 5. *Polyacrylamide gel electrophoresis of nuclear digest DNA. Nuclei were digested as in Fig. 1, isolated DNA was run on a 4% slab gel, stained, photographed and scanned. Migration is from left to right. From bottom to top are: nuclear digests at 2, 4, 9, 18, 32, and 47%, and a chromatin digest at 48% acid soluble DNA. Assigned sizes are shown across the top. (From Sollner-Webb & Felsenfeld, 1975).*

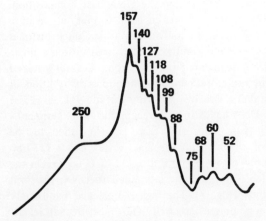

Fig. 6. *Chromatin limit digest DNA. Chromatin was digested to the limit as in Fig. 1. The isolated DNA and sequenced λ hin fragments were run on a 6% polyacrylamide gel. Nuclear digest bands were visualized by staining; hin fragments by autoradiography. Sizes in base pairs are shown above each band. (From Sollner-Webb & Felsenfeld, 1975).*

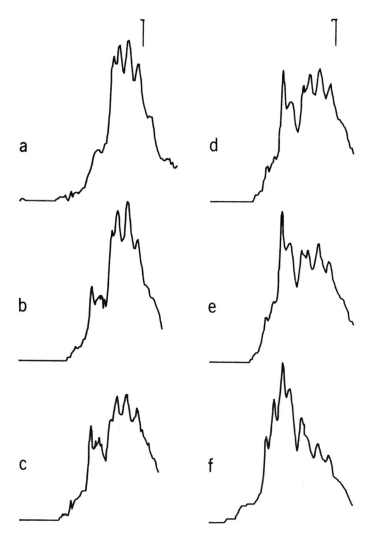

Fig. 7. *Digestion of partial reconstitutes. Electrophoretic patterns of DNA from limit digests of reconstitutes. In each case 0.52 g of H3/H4 were present in the initial complex per gram of DNA. Patterns show the effect on the limit DNA digest of adding to this reconstitution mixture increasing amount of H2A/H2B. The initial ratios of total H2A/H2B to DNA (g/g) were: (a) 0; (b) 0.06; (c) 0.11; (d) 0.17; (e) 0.24; (f) 0.52. (Manuscript in preparation).*

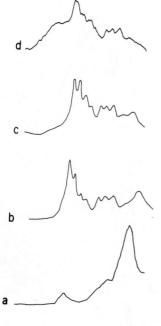

Fig. 8. *Digestion of partial reconstitutes. Electrophoretic patterns of DNA from limit digests of reconstitutes. In each case 0.48 g of H2A/H2B were present per gram of DNA, and varying amounts of H3/H4: (a) 0; (b) 0.12 g; (c) 0.24 g; (d) Limit digest of native chromatin.*

suggest that the upper bands can be generated by combining H2A and H2B with H4, so that these bands may be attributed to the interaction among these three histones. It should be noted that Martinson and McCarthy (1975) have found that H2B and H4 bound to DNA can be crosslinked, but only when H2A is also present.

Like the other histones, the lysine-rich histones H1 and H5 give rise to no discrete DNA bands when reconstituted by themselves. However, they do affect the higher molecular weight bands generated by mixtures of the other histones. If H1 is present in a reconstitution mixture which also contains H2A, H2B, H3 and H4, the upper bands of the digest pattern will split into "doublets" separated by about 3-5 base pairs. The lower bands (arising from H3 and H4) are unaffected.

Our studies lead us to believe that the limit digest bands reflect a highly ordered organization of histones within the beads, which is dependent upon the presence in each bead of more than one histone species. It is possible that, as Kornberg (1974) has suggested, each bead is identical in content, and that the multiplicity of bands we observe arises from internal structural heterogeneity or from the availability of perhaps several degradation pathways. It is also possible that the beads are heterogeneous in histone content, though presumably limited in the range of allowed histone combinations (see, for example, Weintraub (1975)).

III. STRUCTURE IN THE NEIGHBORHOOD OF THE GLOBIN GENE

We now attempt to apply the same kind of chemical probe analysis to the study of protein distributions in the neighborhood of a transcribable gene, namely the globin gene of duck reticulocytes. It was shown in our laboratory some time ago (Axel, Cedar & Felsenfeld, 1973) that chromatin from duck reticulocytes is an active template for *in vivo* transcription of globin message (using *E. coli* RNA polymerase) while chromatin from duck tissues which do not make hemoglobin is not a template for globin message transcription. We have compared globin-specific transcription using chromatin from reticulocytes and erythrocytes. The reticulocyte chromatin is about seven times as active a template as is the erythrocyte chromatin.

In order to study protein distribution in the neighborhood of the globin gene, we make use once again of digestion to the limit with staphylococcal nuclease. We will refer to the DNA of this limit digest as "covered" DNA for convenience. We can also use the chemical probe poly-D-lysine to define a second class of DNA, which we will call "open" DNA. Open DNA is obtained by titrating chromatin with poly-D-lysine until all available sites are occupied. (This requires half as much polylysine as is necessary to saturate the same amount of DNA when free of histones). The titrated chromatin is treated with Pronase to destroy the proteins (but not poly-D-lysine, which is resistant) and then with staphylococcal nuclease to digest the DNA exposed by digestion of the proteins.

Just as in the case of detection of globin mRNA, we can search for globin DNA sequences in open and covered DNA by using radioactively labelled globin cDNA (Ross *et al.*, 1972; Kacian *et al.*, 1972; Verma *et al.*, 1972). When such experiments are carried out with open and covered DNA from duck reticulocytes, we find that the open DNA is missing a discrete portion of the globin probe population, while the covered DNA contains most or all of that population (Fig. 9). When a similar experiment is performed with open and covered DNA from duck erythrocytes, no such difference is seen; the full complement of the globin probe population is present in both.

Detailed analysis of these results (Axel, Cedar & Felsenfeld, 1975) shows that a specific portion of the globin gene in transcriptionally active chromatin is structurally restricted, presumably by protein, in such a way as to prevent reaction with polylysine. The bulk of open and covered DNA, can, however, be shown to be random with respect to sequence population (Axel, Cedar & Felsenfeld, 1975). It seems likely that inactive chromatin possesses the

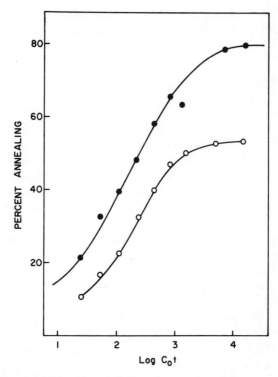

Fig. 9. *Kinetics of annealing of globin cDNA to open and covered region DNA from duck reticulocyte chromatin: 0.1 ng of [^{3}H]-globin DNA was reacted with 5 mg of either covered DNA (●——●) or open region (○——○) DNA from reticulocyte chromatin. Reactions were performed in 0.2 ml and duplex formation was assayed by single strand specific nuclease digestion. (From Axel, Cedar and Felsenfeld, 1975).*

beadlike structure we discussed earlier, and that the beads, although they may be regularly spaced relative to one another, are randomly oriented with respect to base sequence. The transcribable globin gene, on the other hand, appears to have a different organization. It is not yet clear whether the non-random organization of proteins on this gene arises from a different arrangement of histones, the supplementation of bead-and-string structure by non-histone proteins, or the replacement of histones by non-histone proteins.

REFERENCES

Axel, R., Cedar, H., and Felsenfeld, G. (1973). *Proc. Nat. Acad. Sci. U.S.A.* **70**, 2029–2032.
Axel, R., Cedar, H. and Felsenfeld, G. (1975). *Biochem.* **14**, 2489–2495.
Axel, R., Melchior, W., Sollner-Webb, B. and Felsenfeld, G. (1974). *Proc. Nat. Acad. Sci. U.S.A.* **71**, 4101–4105.

Clark, R. J. and Felsenfeld, G. (1971). *Nature, New Biol.* **229**, 101–105.
Clark, R. J. and Felsenfeld, G. (1974). *Biochem.* **13**, 3622–3627.
Hewish, D. and Burgoyne, L. (1973). *Biochem. Biophys. Res. Comm.* **52**, 504–510.
Kacian, D., Spiegelman, S., Bank, A., Terada, M., Metafora, S., Dow, L. and Marks, P. (1972). *Nature, New Biol.* **235**, 167–169.
Kornberg, R. (1974). *Science* **184**, 868–871.
Martinson, H. and McCarthy, B. (1975). *Biochem.* **14**, 1073–1078.
Noll, M. (1974). *Nature* **251**, 249–251.
Olins, A. and Olins, D. (1974). *Science* **181**, 330–332.
Oudet, P., Gross-Bellard, M. and Chambon, P. (1975). *Cell* **4**, 281–300.
Rill, R. and Van Holde, K. (1973). *J. Biol. Chem.* **248**, 1080–1083.
Ross, J., Aviv, H., Scolnick, E. and Leder, P. (1972). *Proc. Nat. Acad. Sci. U.S.A.* **69**, 264–268.
Sollner-Webb, B. and Felsenfeld, G. (1975). *Biochemistry* **14**, 2915–2920.
Verma, I., Temple, G., Far, H. and Baltimore, D. (1972). *Nature, New Biol.* **235**, 163–167.
Weintraub, H. (1975). *Proc. Nat. Acad. Sci. U.S.A.* **72**, 1212–1216.

THE ROLE OF NON HISTONE PROTEINS IN THE REGULATION OF GLOBIN GENE ACTIVITY

R. Stewart Gilmour and A. J. MacGillivray

*Beatson Institute of Cancer Research,
132 Hill Street, Glasgow, Scotland*

I. INTRODUCTION

It may be some time before animal genetics yields information about the molecular mechanisms of eukaryotic gene control in a fashion analagous to its bacterial counterpart. An alternative approach is to isolate the genetic material of animal cells and by manipulating it *in vitro* reveal some of the basic aspects of gene control. In this paper we examine the control of a single gene, the globin gene, in mouse erythroid chromatin. Two important components of the system are considered in detail, namely,
   (a) To show that it is possible to transcribe the globin gene *in vitro* from isolated chromatin and,
   (b) To develop an *in vitro* method for investigating the molecular controls involved.

The discovery that reverse transcriptase from avian myeloblastosis virus can synthesize complementary DNA copies (cDNA) of globin mRNA (Ross et al. 1972; Verma et al. 1972; Kacian et al 1972) offers a hybridization method for analysing the RNA transcribed *in vitro* from chromatin for a specific mRNA sequence. Recently several laboratories (Axel et al. 1973; Gilmour and Paul 1973; Steggles et al. 1974; Gilmour et al. 1974; Barret et al. 1974) have reported that following *in vitro* transcription of a variety of erythroid chromatins globin mRNA sequences can be detected by hybridization to globin cDNA. We have carried out experiments to demonstrate the transcription by *E. coli* RNA polymerase of the globin gene in native chromatin from mouse foetal liver.

In contrast to some of the above reports we find that isolated chromatin contains endogenous globin mRNA sequences which can contaminate the *in vitro* transcripts and give rise to anomalous results. Here it is shown by kinetic and double labeling techniques that globin mRNA sequences are produced *de novo* as a result of polymerase action.

It has also been shown that mouse foetal liver chromatin can be dissociated in 2M NaCl containing 4M urea and then reassociated by gradient dialysis to

0.14M NaCl without affecting the capacity of the chromatin to act as a template for globin mRNA sequences (Gilmour et al. 1973). Furthermore there is evidence to suggest that non-erythroid (mouse brain) chromatin can prime for globin mRNA following reconstitution in the presence of non-histone proteins prepared by hydroxylapatite chromatography of foetal liver chromatin (MacGillivray et al. 1972). In this procedure endogenous RNA tends to cofractionate with non-histone proteins; therefore hybridization analysis of reconstituted chromatin transcripts is open to the same criticism as those from native chromatin. Here we describe a CsCl fractionation method which in conjunction with hydroxylapatite chromatography yields non-histone proteins devoid of endogenous RNA. Chromatin reconstituted from purified DNA, histones and CsCl purified non-histone proteins and transcribed with *E. coli* RNA polymerase clearly produces *de novo* globin mRNA sequences and gives an unambiguous approach to investigating the mechanisms controlling globin gene activity. Using this system it is shown further that a certain sub-fraction of the non-histone proteins from foetal liver regulate globin gene expression.

II. MATERIALS AND METHODS

A. *Preparation of cDNA.*

Globin mRNA was purified from the reticulocytes of mice made anaemic by treatment with phenylhydrazine, by passing polysomal RNA twice through a poly (U)-Sepharose column. The RNA ran as a single 9S component on polyacrylamide gels and directed the synthesis of mouse globin in the duck lysate cell-free system.

A cDNA copy of globin mRNA was prepared with RNA-dependent DNA polymerase from avian myeloblastosis virus as described by Harrison *et al* (1972). [^{3}H] dCTP was incorporated into cDNA to give a final specific activity of 20×10^6 dpm/µg.

B. *Estimation of Globin Sequences in RNA Preparations.*

In this study we used a titration technique in which a fixed amount of cDNA (1ng) is annealed at 43°C with increasing amounts of RNA in 10µl hybridization buffer (0.5M NaCl; 25mM HEPES; 1mM EDTA; 50% formamide, pH 6.7). The theoretical considerations of this approach and the treatment of results are described elsewhere by Young *et al.* (1974). Hybrid was measured by the sensitivity of cDNA to single-stranded nuclease (S_1) prepared from Takadiastase by the method of Sutton (1971). The method is demonstrated in Figure 1 which shows the titration of pure 9S globin mRNA and reticulocyte polysomal RNA to 1 ng globin cDNA. Complete hybridization of the pure globin mRNA is achieved at input ratios of RNA:cDNA = 1.6:1, consistent with the finding that cDNA represents a 60% copy of the mRNA. The input

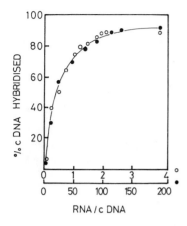

Fig. 1. *Titration to 1ng cDNA of pure 9S globin mRNA (O – O) and reticulocyte polysomal RNA (• – •).*

concentration of polysomal RNA required to achieve saturation is 50 times that needed for pure mRNA, thus giving an estimate of 2% for globin mRNA content. A small percentage of cDNA (5%) is normally resistant to S_1 nuclease in the absence of RNA while an additional 10-15% does not appear to be homologous to 9S RNA and is degraded.

C. Preparation of Chromatin.

Chromatin was prepared by homogenizing tissues in 20 volumes of 1mM Tris HCl (pH 8), 5mM $MgCl_2$, and 0.05mM dithiothreitol (DTT). After 10 min, 0.25 volume of 1M sucrose was added, and the nuclei were sedimented at 2000 x g for 10 min. This step was repeated twice. The resulting pellet was suspended by homogenization in 20 volumes of 2.2M sucrose, 1mM $MgCl_2$ and 0.1mM DTT and centrifuged at 30,000 x g for 60 min. The nuclear pellet was washed successively with 10 volumes homogenizing buffer containing 1% Triton X-100 and 10 volumes 0.14M NaCl. The suspension was sedimented at 2,000 x g for 10 min on each occasion. Finally the pellets were washed with cold distilled water until a clear chromatin gel was obtained.

D. Preparation of High Molecular Weight DNA.

High molecular weight DNA for reconstitution experiments was prepared by the method of Gross-Bellard et al. (1973).

E. Preparation of Chromatin Dependent RNA In Vitro.

RNA was transcribed *in vitro* with *E. coli* RNA polymerase prepared according to the method of Burgess (1969).

Incubations (2 ml) contained 0.04 M Tris (pH 7.9), 2.5mM $MnCl_2$, 1mM DTT, 0.1mM EDTA, 0.8mM each ATP, GTP, CTP and UTP, 100 Burgess units RNA polymerase and 500µg chromatin. After incubation at 37°C for 90 min the incubations were cooled, and 0.1 ml 1 M $MgCl_2$, 0.4 ml 2M KCl and 100µg DNase (Worthington) were added and the reaction incubated at 37°C for 30 min. Pronase (30µl of a 5 mg/ml preincubated solution) was added and the incubation continued for 30 min. 1.5 ml 0.2M EDTA, (pH 7.4), 0.2 ml 10% SDS and 3.7 ml phenol (saturated with 0.5 M KCl) were added, mixed thoroughly and heated at 60°C for 5 min. 3.7 ml chloroform/iso-amyl alcohol (99:1) were added, the mixture heated at 60°C for 5 min and centrifuged at 2,000 x g for 10 min. The organic phase was removed and the remainder re-extracted with chloroform/iso-amyl alcohol at 60°C. The aqueous phase was separated by centrifugation and passed through a Sephadex G50 column in distilled water. Excluded material was collected and dried down.

F. Reconstitution of Chromatin.

Reconstitution experiments were carried out by dissolving native chromatin or mixtures of DNA and chromosomal proteins in 2M NaCl, 4M urea, 0.05 M NaH_2PO_4-Na_2HPO_4 buffer (pH6), 0.01M EDTA, 0.001M DTT, 0.1mM phenylmethylsulphonylfluoride (PMSF) at a concentration of 0 5 mg DNA/ml. The solution was dialyzed against the above medium containing 0.5 M NaCl (16 hrs), and 0.14 M NaCl (2 hrs). The final dialysis was against 0.14 M NaCl alone (2 hrs). The resulting precipitate was pelleted and washed twice in 5 volumes of distilled water to form a gel.

III. RESULTS

A. Transcription of Mouse Foetal Liver Chromatin.

The foetal liver of the mouse embryo reaches maximum haemopoietic activity after 14 days *in utero*, at which time about 70% of the cells are erythroid. Chromatin was prepared from 14 day foetal livers and adult mouse brains and incubated with *E. coli* polymerase. RNA was isolated from a total extraction of the incubation. In a control incubation, mouse foetal liver chromatin was incubated without polymerase and an appropriate amount of *E. coli* soluble RNA added. Figure 2 shows the results obtained when the purified RNAs were hybridized to 1 ng of globin cDNA

RNA transcribed from foetal liver chromatin hybridizes to an extent which suggests that 1 part in 6×10^5 of the RNA represents globin mRNA sequences. RNA transcribed from brain fails to hybridize. The contribution due to endogenous RNA in the foetal liver chromatin was estimated by hybridizing identical aliquots of the control and test incubations. Control incubations contained appreciable amounts of hybridizable RNA; however the corresponding values obtained with RNA synthesized in the presence of polymerase were much higher, suggesting a net increase in globin mRNA sequences arising by *de novo* synthesis.

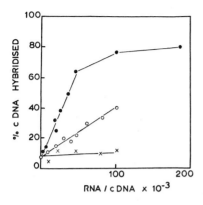

Fig. 2. Titration of RNA transcribed from chromatin with E. coli polymerase, against globin cDNA. Mouse foetal liver chromatin, (● – ●); mouse foetal liver chromatin incubated without polymerase, (○ – ○); mouse brain chromatin, (x – x). In incubations where no polymerase was present E. coli tRNA was added and aliquots hybridized equal to those for complete incubations.

In the absence of polymerase, however, no RNA is synthesized (as judged by [^{3}H] UTP incorporation) and it is concluded that the hybridization observed in control incubations is due to endogenous globin mRNA sequences present in the chromatin. Foetal liver chromatin contains about 10% by weight of endogenous RNA.

In an attempt to distinguish endogenous globin sequences from those arising *de novo*, foetal liver chromatin was incubated in the presence and absence of *E. coli* polymerase for 0, 10, 20, 30 and 40 minutes. Where no polymerase was present *E. coli* soluble RNA was added after half the incubation time had lapsed. The isolated RNAs were hybridized to globin cDNA (Fig. 3). By comparing the titration curves with a standard of pure globin mRNA estimates were made of the total globin mRNA content for each incubation. It is apparent that in the absence of synthesis the net endogenous RNA level falls. Initially the same is true for incubations containing polymerase; however after 30 min there is a net increase in globin sequences over the zero time level.

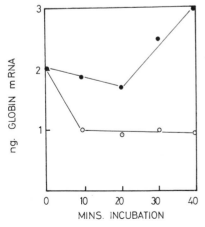

Fig. 3. Kinetic analysis of globin mRNA content of incubations containing foetal liver chromatin in the presence (● – ●) and absence (○ – ○) of E. coli polymerase. Absolute amounts of globin mRNA were calculated for incubations of 0, 10, 20, 30 and 40 min duration by comparison with a standard 9S titration (Fig. 1).

A more reliable control has been employed for measuring both endogenous RNA and newly synthesized RNA simultaneously in the same incubation. RNA was transcribed from foetal liver chromatin as before, incorporating highly labeled [^{32}P]ATP(1-2 Ci/mM). After hybridizing to [^{3}H]cDNA, the reaction mixture was diluted to 0.5 ml with 0.2 M NaCl; 0.05 M Tris-HCl (pH 7.5) and treated with 20 µg/ml pancreatic and T1 ribonucleases at 30°C for 2 hours. The digest was passed through Sephadex G50 in 0.2 M NaCl. The excluded material which contained less than 0.1% of the original ^{32}P counts in RNA was centrifuged to equilibrium in either NaI or CsCl gradients to separate hybridized cDNA from unhybridized cDNA and [^{32}P]RNA. In both cases unhybridized [^{32}P]RNA sediments to the bottom of the gradient. In order to remove as much of this RNA as possible the latter 20% of the gradient was discarded and the remainder containing the [^{3}H]cDNA made up to original volume and re-run. Figure 4 shows the distribution of [^{3}H] and [^{32}P] obtained from the NaI gradient run according to the method of Birnie et al. (1972). [^{3}H] counts appeared at densities of 1.555 and 1.572 corresponding to RNA/DNA hybrid and unhybridized cDNA. [^{32}P] counts were associated with the unhybridized cDNA in amounts from which it could be calculated that [^{32}P] RNA comprised about 40% of the hybridized RNA. In a parallel gradient [^{32}P]RNA was subjected to the same procedure but without hybridization to cDNA. Here no peak of [^{32}P]RNA was found in the hybrid region.

In another experiment the [^{32}P]RNA and [^{3}H]cDNA was analysed on CsCl gradients as described by Szybalski (1968). As can be seen in Figure 5 [^{3}H] counts appeared at densities of 1.78 corresponding to RNA/DNA hybrid and

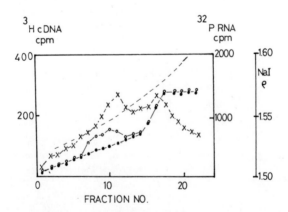

Fig. 4. *Isopycnic banding in NaI of the hybrid formed between [^{3}H] cDNA and [^{32}P] RNA transcribed from mouse foetal liver chromatin. Counts in [^{3}H] (x – x); counts in [^{32}P], (○ – ○); density (- - - -). A sample of [^{32}P] RNA treated in an identical fashion but without hybridization to cDNA was run in a separate gradient, (● – ●).*

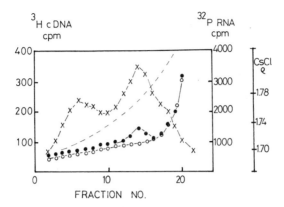

Fig. 5. *Isopycnic banding in CsCl of the hybrid formed between [^3H] cDNA and [^{32}P] RNA transcribed from mouse foetal liver chromatin. Counts in [^3H], (x – x); counts in [^{32}P], (● – ●); density (dashed line). A sample of [^{32}P] RNA treated in an identical fashion but without hybridization to cDNA was run in a separate gradient, (○ – ○).*

1.71 corresponding to unhybridized cDNA. From the [^{32}P] counts associated with the hybridized cDNA it was calculated that [^{32}P] accounted for 60% of the hybridized RNA.

B. Reconstitution Experiments.

Reconstitution experiments were devised to test whether the specific transcription of the globin gene in erythroid chromatin is an inherent property of the chromatin and if so which structural elements are responsible for directing this specificity.

In previous work (Gilmour 1975) the functional integrity of foetal liver chromatin was investigated by dissolving native chromatin in 2M NaCl, 5M urea, 0.01M Tris-HCl, pH8.0. This procedure dissociates the DNA and protein components of chromatin. The nucleoprotein complex was then reconstituted by gradient dialysis as described in Materials and Methods. RNA transcribed from reconstituted chromatin was compared with that from undissociated chromatin by hybridization to globin cDNA. Control incubations contained chromatin but no polymerase. The data in Figure 6 show that RNA transcribed from both chromatins hybridized to about the same extent; however it can be seen from control incubations that the reconstitution process does not eliminate the background due to endogenous RNA. Recent modifications of this procedure now make it possible to reconstitute chromatin with the exclusion of endogenous RNA. Foetal liver chromatin was prepared as far as the 0.14M NaCl wash. Pelleted material was dissolved in 2M NaCl, 4M urea, 0.05M Na_2PO_4-Na_2HPO_4 buffer, (pH6), 0.01M EDTA, 0.001M DTT, 0.1mM PMSF

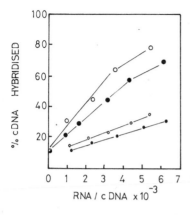

Fig. 6. Titration against globin cDNA of RNA transcribed from native mouse foetal liver chromatin, (O – O); reconstituted mouse foetal liver chromatin, (● – ●); RNA from incubations containing native, (○ – ○), and reconstituted chromatins (● – ●) and no polymerase.

as described in Materials and Methods. An equal volume of 55% (w/v) caesium chloride, 4M urea; 0.05M NaH_2PO_4-Na_2HPO_4 buffer, (pH6), 0.01M EDTA, 0.001M DTT was added to give a final DNA concentration of 0.3-0.5 mg/ml. 4.5 ml of dissolved chromatin were centrifuged in an MSE 10 x 10 ml. titanium rotor (the remainder of the tube was filled with light paraffin) at 40,000 rpm for 40 hrs at 8°C. Fractions (0.5ml) of the gradient were collected and monitored at 280nm. The chromatin proteins are found in the top 1.5 ml of the gradient while the DNA and RNA form a pellet.

The chromosomal proteins comprising histones and non-histones were mixed with purified DNA in 1.5:1 (w/w) proportions and dialysed against 2M NaCl, 4M urea, 0.05M NaH_2PO_4-Na_2HPO_4 buffer (pH6), 0.01M EDTA, 0.001M DTT, 0.1mM PMSF for 2 hrs, then against the same buffer containing 0.5M NaCl and 0.14M NaCl for 16 hrs and 2 hrs respectively. Finally the chromatin was dialysed against 0.14M NaCl alone at which point precipitation occurred. The reconstituted chromatin was pelleted and washed twice in distilled water to form a gel.

RNA transcribed from foetal liver reconstituted chromatin in the presence of E. coli polymerase was hybridized to globin cDNA (Fig. 7). A control incubation without polymerase was carried out as before. The results show that little, if any endogenous RNA is present in the reconstituted chromatin. In the presence of polymerase *de novo* globin mRNA sequences are transcribed.

This modified reconstitution technique was extended to show that the non-histone component of foetal liver chromatin is specifically required for the expression of globin genes in the reconstituted chromatin. Chromsomal proteins from both foetal liver and mouse brain were prepared on CsCl-urea gradients. The histones and non-histones from each were separated by chromatography on hydroxylapatite by the method of MacGillivray et al (1972). The histone fractions were pooled. Samples of non-histone protein from foetal liver or brain were reconstituted with purified mouse DNA and pooled histones. (In both cases

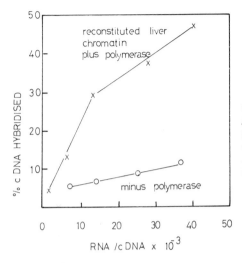

Fig. 7. *Titration against globin cDNA of RNA transcribed from mouse foetal liver chromatin reconstituted from purified mouse DNA and CsCl purified chromatin protein in the presence (X – X) and absence (O – O) of E. coli polymerase.*

protein:DNA = 1.5:1). The chromatins were reconstituted and transcribed. The data in Figure 8 show the hybridization of the isolated RNAs to globin cDNA. It can be seen that the globin gene is available for transcription only when foetal liver proteins are present in the reconstitution mixture. Chromatin reconstituted with brain non-histone proteins fails to hybridize. The background obtained with reconstituted foetal liver chromatin in the absence of polymerase is only slightly above this level.

Preliminary experiments have been carried out to show that the control of globin gene transcription is a function of a particular subfraction of foetal liver non-histone proteins. Foetal liver chromatin was dissociated in 2M NaCl, 4M urea, 0.01M EDTA, 0.001M DTT, 0.1mM PMSF and applied to a hydroxylapatite column as described by Rickwood & MacGillivray (1975). Various fractions of chromosomal proteins were obtained according to the

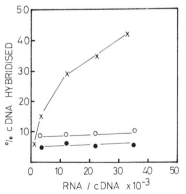

Fig. 8. *Titration to cDNA of RNA transcribed from chromatin reconstituted from a common pool of mouse DNA and histones and CsCl purified non-histone proteins from mouse foetal liver (x – x) and mouse brain (• – •). A control incubation was set up containing reconstituted mouse foetal liver chromatin, E. coli soluble RNA and no polymerase (○ – ○).*

elution sequence described in Table I. Fraction H_1 comprises histones while H_2-H_4 contain non-histone proteins with increasing affinity for hydroxylapatite. The individual fractions were then diluted with an equal volume of 55% CsCl, 4M urea and purified as described above. Purified mouse embryo DNA was then reconstituted with H_1 alone, with H_1 in combination with either H_2, H_3 or H_4 separately and with H_1, H_2, H_3 and H_4 together. In each case a 1.5 fold excess of protein fractions were reconstituted. The RNA transcribed from the resulting chromatins were hybridized to globin cDNA as shown in Figure 9. Most of the protein responsible for comparing specific transcription of the globin gene is found in the H_2 fraction with some overlap into fraction H_3. Proteins from H_4 did not reconstitute a chromatin with globin gene activity.

IV. DISCUSSION

In a previous publication we reported that *in vitro* transcripts of foetal liver chromatin contained little endogenous globin mRNA sequences (Gilmour and Paul 1973). Both Axel et al. (1973) and Steggles et al. (1974) reported similar findings with different erythropoietic systems. We have re-examined the question of a suitable control for chromatin primed reactions in view of the

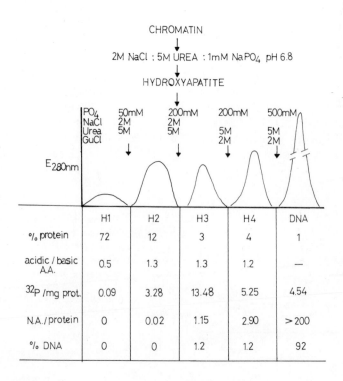

	H1	H2	H3	H4	DNA
% protein	72	12	3	4	1
acidic / basic A.A.	0.5	1.3	1.3	1.2	—
^{32}P /mg prot.	0.09	3.28	13.48	5.25	4.54
N.A./protein	0	0.02	1.15	2.90	>200
% DNA	0	0	1.2	1.2	92

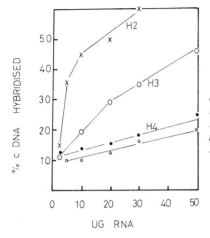

Fig. 9. *Titration of cDNA of RNA transcribed from chromatin reconstituted from mouse DNA and histones and non-histone protein fractions from hydroxylapatite chromatography H_2 (x – x), H_3 (O – O), H_4 (● – ●) and controls containing no polymerase (○ – ○).*

finding that when carrier RNA is present in enzyme minus control incubations substantial amounts of endogenous globin mRNA sequences can be detected in the total RNA extract (Gilmour et al. 1975). Barrett et al. (1974) have also reported endogenous globin mRNA in chick reticulocyte chromatin.

The reason for these discrepancies is not clear. It can be seen from the data in Figure 3 that the amount of endogenous globin mRNA in control incubations decreases with time. This is possibly due to the presence of small amounts of nuclease activity in chromatin and it is probable that this property varies between different chromatin systems. Here we have attempted to distinguish clearly between the endogenous mRNA sequences in foetal liver chromatin and those produced by *in vitro* transcription with E. coli polymerase. While it is possible to show a net increase in the total globin mRNA sequences in the presence of polymerase (Fig. 3), the question still remains as to whether a "minus enzyme" incubation gives a true estimate of endogenous contamination. The double label experiments shown in Figures 4 and 5 provide the least ambiguous answer. In both instances it was found that 50% or more of the hybridizing globin mRNA could be attributed to newly transcribed sequences. These controls confirm the conclusion that the chromatins of differentiated erythroid tissue have active globin genes while in other tissues this gene is repressed (Fig. 1).

The functional reconstitution of chromatin offers a method of investigating the underlying mechanisms of this differential gene expression. However it is apparent that endogenous RNA is still present after a simple one-step reconstitution procedure (Fig. 6). From the point of view of employing reconstitution as a routine assay for the activity of gene regulators, interpretation of the data would be greatly facilitated if RNA-free preparations were available.

It has proved possible to do this by a two-step reconstitution procedure in which the chromosomal proteins are initially dissociated and separated free of nucleic acid by CsCl-urea gradients. Reconstitution of CsCl purified proteins to purified DNA yields a functionally active chromatin that is essentially devoid of endogenous RNA (Fig. 7). Thus it is possible to confirm the conclusions of Figure 1 in a more decisive manner. By fractionating purified chromosomal proteins on hydroxylapatite and reconstituting them individually to purified DNA (Fig. 8) it can be shown, without recourse to elaborate controls, that the activity of the globin gene is under direct influence of the non-histone component of erythroid chromatin. In agreement with this finding Barrett et al. (1974) have recently used a urea-guanidinium chloride centrifugation step to prepare RNA-free non-histone proteins from chick reticulocytes, and have shown that this fraction can activate the globin genes in reconstituted erythrocyte chromatin.

We have applied these techniques to assay for globin gene activation in a number of subfractions of foetal liver non-histone proteins obtained by hydroxylapatite chromatography. The bulk of the activity appears in the fraction of protein that elutes in 0.05M sodium phosphate with some overlap into the H_3 fraction (Fig. 9). Chromatins reconstituted with histones and any one of the non-histone proteins are found to support RNA synthesis. While the RNA synthesized from H_2 reconstituted chromatin contains about the same proportion of globin mRNA sequences as that from native chromatin that from H_3 reconstituted chromatin was considerably less. Transcripts from H_4 reconstituted chromatin contained no globin mRNA sequences. We interpret the intermediate response as being due to either an insufficiency of specific non-histone proteins in H_3 to activate all globin genes present in the DNA or to a proportionately greater synthesis of RNA from activated genes other than those coding for globin. This experiment suggests that globin gene activation is a property of only certain non-histone protein species. The H_2 hydroxylapatite proteins are still quite heterogeneous as judged by two dimensional gels, and it is not yet clear whether a single species is responsible or whether several different species act co-operatively. We are investigating this possibility in current experiments by sequentially fractionating the H_2 proteins into smaller, less complex groups and testing each for globin gene activation.

REFERENCES

Axel, R., Cedar, H. and Felsenfeld, G.w(1973). *Proc. Nat. Acad. Sci. U.S.A.* **70**, 2029–2032.

Barrett, T., Maryanka, D., Hamlyn, P. H. and Gould, H. J. (1974). *Proc. Nat. Acad. Sci. U.S.A.* **71**, 5057–5061.

Burgess, R. R. (1969). *J. Biol. Chem.* **244**, 6160–6167.

Gilmour, R. S. and Paul, J. (1973). *Proc. Nat. Acad. Sci. U.S.A.* **70**, 3440–3442.

Gilmour, R. S., Humphries, S. E., Hale, E. C. and Paul, J. (1973). *In* "Normal and Pathological Protein Synthesis", p. 89. INSERUM Colloquivm, Paris.

Gilmour, R. S., Windass, J. D., Affaro, N. and Paul, J. (1975). *J. Cell Physiol.* **85**, 449–458.

Gross-Bellard, M., Oudet, P. and Chambon, P. (1973). *Eur. J. Biochem.* **36**, 32–38.

Harrison, P. R., Hell, A. and Paul, J. (1972). *FEBS Letters* **24**, 73–76.

Kacian, D. L., Spiegelman, S., Bank, A., Terada, M., Metaphora, S., Dow, L. and Marks, P. A. (1972). *Nature New Biol.* **235**, 167–169.

MacGillivray, A. J., Cameron, A., Krauze, R. J., Richwood, D. and Paul, J. (1972). *Biochim. Biophys. Acta* **277**, 384–402.

Rickwood, D. and MacGillivray, A. J. (1975). *Eur. J. Biochem.* **51**, 593–601.

Ross, J., Aviv, A., Scolnick, C. and Leder, P. (1972). *Proc. Nat. Acad. Sci. U.S.A.* **69**, 264–268.

Steggler, A. W., Wilson, G. H., Kantor, J. A., Picciano, D. J., Falvey, A. K. and Anderson, W. F. (1974). *Proc. Nat. Acad. Sci. U.S.A.* **71**, 1219–1223.

Sutton, W. D. (1971). *Biochim. Biophys. Acta* **240**, 522–531.

Szybalski, W. (1968). *In* "Methods in Enzymology" (S. P. Colowick and N. O. Kaplan, eds.), Vol. 12B, p. 330. Academic Press, New York.

Verma, I. M., Temple, G. F., Fan, H. and Baltimore, D. (1972). *Nature New Biol.* **235**, 163–167.

Young, B. D., Harrison, P. R., Gilmour, R. S., Birnie, G. D., Hell, A., Humphries, S. and Paul, J. (1974). *J. Mol. Biol.* **84**, 555–568.

TABLE I.
General Properties of Animal Cell RNA Polymerases

	Enzyme Class			
	I	II	III	
Elution Position From DEAE-Sephadex	0.1	0.2	0.2-0.3	$M\ (NH_4)_2SO_4$
Elution Position from DEAE-Cellulose	0.1	0.2	0.1	$M\ (NH_4)_2SO_4$
Ionic Strength Optima	0.05	0.10	0.05 and 0.16	$M\ (NH_4)_2SO_4$
Activity Ratio Mn^{++}/Mg^{++}	1	5-10	2	
Activity Ratio poly(dA-dT)/DNA	1-2	0.5-1	5-10	
α-Amanitin Sensitivity	none	extreme	intermediate	

These data summarize the properties observed for the RNA polymerases from *Xenopus laevis* oocytes and somatic cells (Roeder, 1974; Roeder, unpublished observations), from mouse plasmacytoma cells (Schwartz et al., 1974), and from human KB cells (J. Jaehning, unpublished observations).

defined conditions (Roeder and Rutter, 1969; Roeder, 1974a; Schwartz et al., 1974a). The class I, II, and III RNA polymerases are readily distinguished by several of these properties including 1) elution positions from DEAE-Sephadex, 2) ionic strength optima, 3) relative activities with Mn^{++} versus Mg^{++}, 4) relative activities with synthetic templates versus native DNA (see also Blatti et al., 1970), and 5) α-amanitin sensitivities. Since the α-amanitin sensitivities are of special significance for functional studies (Section IV) they are shown in more detail in Figure 1. The mouse plasmacytoma RNA polymerases II and III both show complete sensitivity to α-amanitin although the former is about 10^3 fold more sensitive than the latter (upper panel). In contrast, the murine RNA polymerase I is completely insensitive to high concentrations of this toxin. The class I, II, and III enzymes isolated from oocytes of *X. laevis* show α-amanitin sensitivities identical to those of the murine enzymes (center panel). Significantly, the class I, II, and III activities are readily distinguished in mixtures of the enzyme. A low toxin concentration selectivity inhibits RNA polymerase II and a high toxin concentration selectively inhibits both RNA polymerases II and III (Weinmann and Roeder, 1974; Weinmann et al., 1974; Schwartz et al., 1974a).

The distinctive properties of the enzymes from animal cells suggest that the class I, II, and III enzymes differ in structure and in function (Sections IV and V). However, none of the chromatographically distinct enzyme forms within a

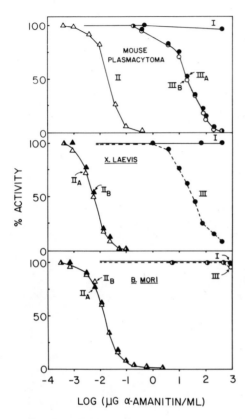

Fig. 1. *Sensitivity of eukaryotic RNA polymerases to α-amanitin. Upper panel, murine plasmacytoma enzymes. Center panel, Xenopus laevis oocyte enzymes. Lower panel, Bombyx mori posterior silk gland enzymes. The murine enzymes were solubilized from solid MOPC 315 tumors, the X. laevis enzymes from adult ovaries containing predominantly mature oocytes, and the B. mori enzymes from posterior silk glands taken from fifth instar larvae (day 5). In each case the various class I, II, and III enzymes indicated were resolved by chromatography on DEAE-Sephadex. Each of the class I enzyme preparations analyzed contain distinct enzyme forms (I_A and I_B) which can be resolved chromatographically (X. laevis and B. mori) or electrophoretically (murine plasmacytoma, see text). The murine RNA polymerase II preparation contains forms II_O, II_A, and II_B which can be resolved electrophoretically (see text). The class III enzyme preparations represent in each case single chromatographic species (see text). The partially purified enzymes were assayed with calf thymus DNA as template in the presence of varying concentrations of α-amanitin. For details see Schwartz et al. (1974a) and Roeder (1974a).*

given class can be distinguished on the basis of catalytic properties or sensitivities to α-amanitin as shown by analysis of forms I_A and I_B from *X. laevis*, forms II_A and II_B from *X. laevis*, and forms III_A and III_B from mouse myeloma cells (Roeder, 1974a; Schwartz *et al.*, 1974a; Fig. 1). These latter findings support the grouping of the various forms within specific classes and further suggest that the enzymes within a class may possess only minor differences in structure and that they may perform related functions.

Many of the properties listed in Table I are also similar for the analogous Class I, II, and III enzymes from a variety of other eukaryotes including yeast (Adman *et al.*, 1972), sea urchins (Roeder and Rutter, 1969), the silk moth *Bombyx mori* (R. G. Roeder and L. P. Gage, unpublished observations), and human cells (Sergeant *et al.*, 1974; Jaehning, unpublished observations), suggesting that basic structural, functional, and regulatory features of these enzymes have been retained throughout evolution. However, it may be noted that in some organisms one or more of the properties of a given RNA polymerase may differ somewhat from those indicated in Table I (for references see Chambon, 1975). Since α-amanitin is conveniently used to investigate enzyme function it is particularly relevant to note that the sensitivities of the RNA polymerases to this toxin differ widely among various organisms and, therefore, cannot always be used to distinguish RNA polymerases I, II, and III. For example, as shown in Figure 1 (lower panel), both RNA polymerases I and III of the silk moth *Bombyx mori* are resistant to the high concentrations of α-amanitin which completely inhibit the murine enzyme III, even though the *B. mori* class II enzymes show sensitivities similar to those of the murine RNA polymerase II. In yeast, class I and III enzymes are both resistant to high concentrations of α-amanitin and RNA polymerase II shows only an intermediate sensitivity (Adman *et al.*, 1972; L. Schultz, personal communication). Such differences between the analogous enzymes from various eukaryotes are likely to reflect structural and possibly functional differences. However, many other properties (including the first five listed in Table I) of the yeast (Adman *et al.*, 1972) and the *Bombyx mori* (unpublished observations) class I, II, and III RNA polymerases closely resemble those of the analogous enzymes of animal cells, suggesting that at least some of the structural and functional properties of the analogous enzymes in these various organisms are the same.

IV. GENERAL FUNCTIONS OF CLASS I, II, AND III RNA POLYMERASES

A. *Transcription of Cellular Genes.*

The class I, II, and III RNA polymerases from animal cells have been shown to have distinct cellular functions in agreement with the observations that they have distinct properties (Section III) and distinct subunit structures (Section V). These experiments have taken advantage of the findings that isolated nuclei are active in the synthesis of defined classes of RNA *in vitro* and that the endogenous RNA polymerase activities in these nuclei show the same sensitivities to α-amanitin as do the purified enzymes (Lindell *et al.*, 1970; Weinmann and Roeder, 1974; Weinmann, Raskas and Roeder, 1974). Specific RNA species synthesized in nuclei have been identified either electrophoretically or by DNA-RNA hybridization competition techniques. The general functions which have been established for the animal cell enzymes in the transcription of cellular genes are summarized in Table II. RNA polymerase I is involved in the synthesis of the 18s and 28s rRNA's and RNA polymerase II is involved in HnRNA synthesis (Blatti *et al.*, 1970; Zylber and Penman, 1971; Reeder and Roeder, 1972; Weinmann and Roeder, 1974). Since HnRNA is thought to contain precursors for mRNAs (Darnell, Jelinek, and Molloy, 1973) it is probable that mRNA synthesis is also mediated by RNA polymerase II although this has not been demonstrated directly. Synthesis of the low molecular weight tRNA precursor and 5S rRNA species is mediated by RNA polymerase III (Weinmann and Roeder, 1974; Weinmann et al., 1975). These assignments of enzyme function are consistent with the known subnuclear localizations of the class I, II, and III RNA polymerases (Roeder and Rutter, 1970; Schwartz et al., 1974) as summarized in Table II. Varying amounts of the RNA polymerases are found in the cytoplasmic fraction following subcellular fractionation (Schwartz et al., 1974a; Seifart, Benecke, and Juhasz, 1973; Seifart and Benecke, 1975). However, this is not necessarily indicative of either storage or function of the enzyme(s) in the cytoplasm of the living cell because of the possibility of nuclear leakage during isolation (for discussion of this problem see Schwartz et al., 1974a).

B. Transcription of Viral Genes.

In addition, the functions of the class I, II, and III enzymes have been examined during the lytic infection of human KB cells by adenovirus 2. The adenovirus 2 genome is a linear double stranded DNA of about 23×10^6 daltons. During the lytic cycle there are qualitative and quantitative changes in the rates of viral gene transcription (for references see Green, 1970 Cold Spring Harbor Symposium, 1974; and Weinmann, Jaehning, Raskas and Roeder, 1975). Two discrete classes of viral RNA can readily be distinguished during late periods of viral infection (after DNA replication). The first class of viral transcripts consists of mRNAs (or mRNA precursors) which are transcribed from greater than half the viral genome (Sharp, Gallimore, and Flint, 1974; Green, 1970) and which account for the bulk of viral RNA synthesis. The second class of viral transcripts

TABLE II.
Localization and General Functions of Animal Cell RNA Polymerases

Enzyme Class	Subnuclear Localization	Cellular Gene Transcripts	Viral Gene Transcripts
I	nucleolus	18S, 28S rRNA's	none identified
II	nucleoplasm	HnRNA's	mRNA precursors
III	nucleoplasm	tRNA's, 5S rRNA	low molecular weight RNA's (V_{200}, V_{156}, V_{140})

The functions of the RNA polymerases in transcribing cellular genes were analyzed in mouse plasmacytoma cells, in amphibian cells, and HeLa cells. The functions of the RNA polymerases in transcribing viral genes were analyzed in adenovirus 2 infected KB cells. The viral V_{200}, V_{156}, and V_{140} species are approximately 200, 156, and 140 nucleotides in length, respectively. For references and discussion, see text.

is comprised of low molecular weight RNAs which are synthesized in large molar amounts but whose functions remain undefined. The most prominent RNA in this class is the viral 5.5S RNA which is 156 nucleotides in length and which has been well characterized by Ohe and Weissman (1971). More recently two new viral coded RNA's containing, respectively, 200 and 140 nucleotides have been identified (Weinmann et al., 1975) in lytically infected cells. These small viral RNA's are designated here as V_{200} (200 nucleotides), V_{156} (156 nucleotides) and V_{140} (140 nucleotides).

Nuclei isolated from adenovirus infected cells are extremely active in the synthesis of viral mRNA's (or their precursors) (Wallace and Kates, 1972; Price and Penman, 1972a; Weinmann et al., 1974, 1975b) as well as the V_{200}, V_{156}, and V_{140} viral RNA species (Weinmann et al., 1974, 1975). Using α-amanitin to inhibit the endogenous RNA polymerase activities in nuclei from infected cells, it has been clearly shown that viral mRNA synthesis (monitored by hybridization to viral DNA) is mediated by an RNA polymerase II activity whereas the three low molecular weight viral RNAs (monitored electrophoretically) are synthesized by an RNA polymerase III activity (Weinmann et al., 1974; Weinmann et al., 1975b). No involvement of RNA polymerase I in viral gene transcription could be demonstrated These data are summarized in Table II.

These findings provide more convincing support for the idea that RNA polymerase II is involved directly in the synthesis of cellular mRNA or mRNA precursors while the RNA polymerase III is involved in the synthesis of low molecular weight cellular RNA's. Furthermore, fragments of DNA containing the sequences encoding the small viral RNA's (V_{200}, V_{156}, and V_{140}) have been mapped on the viral genome using the restriction endonuclease digestion technique. They appear to be clustered within 0.18-0.34 fractional units of the adenovirus DNA and are all transcribed in the same direction (Weinmann et al., 1975b; King, Raskas and Roeder, unpublished observations). While the significance of this finding is as yet unclear, it is relevant to note that the reiterated cellular genes transcribed by class III enzymes are also arranged in clusters (Brown, Wensink, and Jordon, 1971; Clarkson and Birnsteil, 1973). It is tempting to speculate that the reiterated genes (Marsluff et al, 1975) encoding the small nuclear RNA's found in eukaryotic cells (Weinberg and Penman, 1969) are also transcribed by class III RNA polymerases.

These studies of cellular and viral gene transcription clearly establish that distinct classes of genes are transcribed by distinct classes of RNA polymerases. It is also probable that the enzymes within a class have distinct but related functions. However, this question is currently unresolved since it has not yet been possible to functionally distinguish the activities of the enzymes within a given class (e.g., with α-amanitin). Reconstruction experiments with separated enzymes, defined DNA or nucleoprotein templates, and specific assay systems will likely be useful in investigating this problem.

V. MOLECULAR STRUCTURES OF CLASS I, II, AND III RNA POLYMERASES

A. Resolution of Multiple Forms.

Studies of RNA polymerase structure represent an important area of investigation since the structural complexities and relationships of the various enzymes should provide some insight into potential means of regulation of the various enzyme concentrations and activities. The early studies of RNA polymerase II from calf thymus and rat liver (Weaver et al., 1971; Kedinger and Chambon, 1972) suggested that the animal RNA polymerases were multisubunit enzymes, with complexities equal to or greater than the bacterial enzymes (Burgess, 1971). Later studies of the calf thymus RNA polymerase I (Gissinger and Chambon, 1972) revealed a similar complexity for this enzyme although most of the subunits appeared distinct from those of RNA polymerase II. More recently our laboratory has focussed on structural studies of the RNA polymerases in mouse plasmacytoma cells which have high levels of class I, II, and III RNA polymerases, thereby permitting analysis of the three major classes of the enzymes in a single cell type.

When RNA polymerase extracts from mouse plasmacytoma cells are subjected to chromatography on DEAE-Sephadex, the major enzyme forms I and II and two minor enzyme forms designated III_A and III_B are resolved (Schwartz et al., 1974a). Each of these enzymes has been purified to apparent homogeneity by ion exchange chromatography and sucrose gradient sedimentation (Schwartz and Roeder, 1974, 1975; Sklar and Roeder, 1975). Although chromatographic heterogeneity has been demonstrated in RNA polymerases I and II from other organisms (section II) it has not been possible to chromatographically resolve individual class I or class II mouse plasmacytoma enzymes. To investigate the possibility that the chromatographically purified enzymes, in fact, represent mixtures of enzymes, each preparation was subjected to polyacrylamide gel electrophoresis under non-denaturing conditions. As shown by the optical density scans of the stained polyacrylamide gels (Fig. 2) each enzyme exhibits two (I, III_A, and III_B) or three (II) major bands. For each enzyme the subunit composition of the individual electrophoretic bands have been analyzed via subsequent electrophoresis in the presence of SDS.

B. Class I RNA Polymerases.

The polypeptide compositions of the two electrophoretic forms of RNA polymerase I (Fig. 2) as well as that of the chromatographically purified (phosphocellulose) enzyme are shown in Figure 3. The electrophoretic form I_A contains stoichiometric amounts of six polypeptides designated subunits Ia, Ib, Ic, Id, Ie, and If. The molecular weights of these components are respectively, 195,000, 117,000, 61,000, 52,000, 29,000 and 19,000. The polypeptide

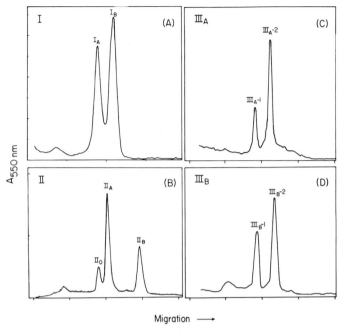

Fig. 2. *Polyacrylamide gel electrophoresis of mouse plasmacytoma RNA polymerases under non-denaturing conditions. RNA polymerases I, II, III$_A$, and III$_B$ (Schwartz et al., 1974a) from MOPC 315 tumors were purified by ion exchange chromatography (DEAE-Sephadex, DEAE-cellulose, CM-Sephadex, phosphocellulose) and by sucrose gradient sedimentation. Samples of each chromatographically purified enzyme were subjected to electrophoresis on 5% polyacrylamide gels under non-denaturing conditions. After electrophoresis, gels were stained for protein, destained, and scanned at 550 nm. The scan of the rightward 40-50% of each gel is not shown since no protein was detected in these regions of the gel. For details of enzyme purification and gel electrophoresis see Schwartz and Roeder, 1974; 1975; and Sklar and Roeder, 1975.*

composition of form I_B is identical except for the absence of the 61,000 dalton polypeptide (subunit Ic). These observations suggest the possibility of two functionally distinct forms of RNA polymerase I within the cell. These could represent enzymes which transcribe distinct genes or, alternatively, distinct functional states (e.g., active versus inactive) of an enzyme which transcribes a single class of genes. In either case it is also possible that a modified Ic subunit or a distinct polypeptide of similar size is normally associated with enzyme I_B and that this polypeptide is dissociated from the enzyme during electrophoresis under non-denaturing conditions. This is suggested by direct analysis of the polypeptide composition of the chromatographically purified (phosphocellulose) enzyme. As shown in the upper panel of Figure 3 this preparation appears to contain equimolar or greater amounts of either the 61,000 dalton polypeptide(s)

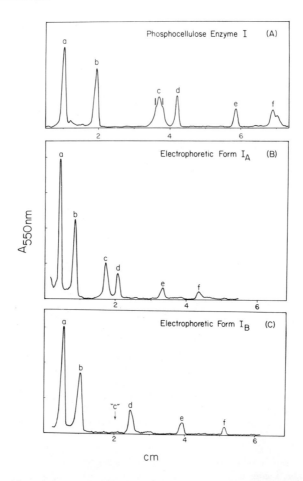

Fig. 3. *Polypeptide compositions of the chromatographic and electrophoretic forms of murine plasmacytoma RNA polymerase I. Panel A, chromatographically purified RNA polymerase I (phosphocellulose fraction). Panel B, electrophoretic form I_A. Panel C, electrophoretic form I_B. The electrophoretic forms I_A and I_B were resolved by electrophoresis of chromatographically purified RNA polymerase I under non-denaturing conditions as described in Figure 2. Samples of each enzyme were subjected to electrophoresis on 12% polyacrylamide gels in the presence of SDS. After electrophoresis gels were stained for protein, destained, and scanned at 550 nm. For details of the electrophoretic techniques see Schwartz and Roeder, 1974.*

or associated polypeptides of similar size (indicated by the vertical bars). Despite these suggestions of structural heterogeneity in RNA polymerase I, however, there is as yet no convincing evidence that this reflects functional heterogeneity within the cell.

C. Class II RNA Polymerases.

The polypeptide compositions of the three electrophoretic forms of RNA polymerase II (Fig. 2) are shown in Figure 4. Panels A, B, and C show the results obtained with electrophoretic forms II_O, II_A and II_B, respectively. Each form contains equivalent molar amounts of subunits IIc, IId, IIe, IIf, IIg, IIh, and IIi which have approximate molecular weights, respectively, of 140,000, 41,000, 29,000, 27,000, 22,000, 19 000, and 16,000. The distinguishing feature of each enzyme form rests with the high molecular weight subunits. The 170 000 dalton subunit IIb is unique to form II_B, the 205,000 dalton subunit IIa is unique to form II_A and the 240,000 dalton subunit IIo is unique to form II_O. The presence of subunit IIa in the II_O sample analyzed in Figure 4 is due to incomplete separation of form II_O from form II_A as revealed by subsequent experiments (Schwartz and Roeder, 1975). For each enzyme form the molar ratios of each subunit, relative to subunit IIc, are near unity with the exception of subunit IIh which has a ratio of about 2. More recent studies reveal that subunit IIh is composed of two distinct subunits (IIh1 and IIh2) present in equimolar ratios (Sklar et al., 1975).

These studies reveal at least three structurally distinct forms of RNA polymerase II. It may be speculated that these enzyme forms have distinct functions in the cell, although this has not yet been demonstrated. The class II enzymes from mouse plasmacytoma are also considerably more complex than those reported for the analogous enzymes from other animal cell types (Weaver et al., 1971; Kedinger and Chambon, 1972; Sugden and Keller, 1973) although in the present studies the enzyme was obtained in a soluble form without the normal high salt solubilization technique and the enzymes were purified with somewhat different procedures. This resulted in substantially higher yields of enzyme activity (Schwartz and Roeder, 1975) than reported previously and may also account for the presence of additional enzyme associated polypeptides. More recent studies in other organisms also suggest structural complexities for class II enzymes analogous to those reported here (Kedinger et al., 1974).

D. Class III RNA Polymerases.

Further analysis of the electrophoretic forms of RNA polymerase III_A (Fig. 2) is shown in Figure 5 (panels B and C). When forms III_A-1 and III_A-2 are subjected to electrophoresis under denaturing conditions ten distinct polypeptides are evident. These polypeptides are designated subunits IIIa, IIIb, IIIc, IIId, IIIe1, IIIe2, IIIf, IIIg, IIIh, and IIIi and have molecular weights respectively, of 155,000, 138,000 89,000, 70,000, 53 000, 49,000, 41,000, 32,000, 29,000, and 19,000. Molar ratios are close to unity for all subunits except for IIIi, which is present in stoichiometric excess. These same polypeptides are present in both electrophoretic forms of III_A (panels B and C) as well as in the chromatog-

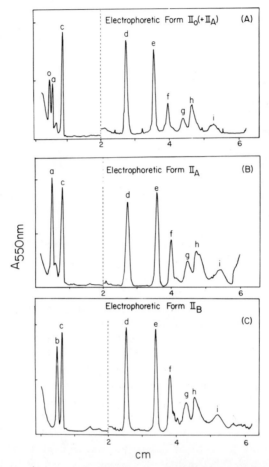

Fig. 4. *Polypeptide compositions of the electrophoretic forms of murine plasmacytoma RNA polymerase II. Panel A, electrophoretic form II_O (containing in addition a significant amount of form II_A, see text). Panel B, electrophoretic form II_A. Panel C, electrophoretic form II_B. A chromatographically purified preparation of RNA polymerase II was subjected to electrophoresis under non-denaturing conditions as described in Figure 2. Individual electrophoretic forms were subsequently subjected to electrophoresis in 12% polyacrylamide gels containing SDS. Gels were stained for protein, destained, and scanned at 550 nm. In each panel the optical density scales for the 2-6 cm regions of the gels were expanded approximately 3 fold relative to the scales for the 0-2 cm regions of the gels, in order to better visualize the low molecular weight polypeptide profiles. For details of the electrophoresis procedures see Schwartz and Roeder, 1975.*

raphically purified (phosphocellulose) enzyme III_A (panel A). Furthermore, the stoichiometry of these polypeptides is the same in each case, with the possible exception of subunit IIIf which shows some heterogeneity within each of the electrophoretic forms. Under some conditions of electrophoresis, heterogeneity

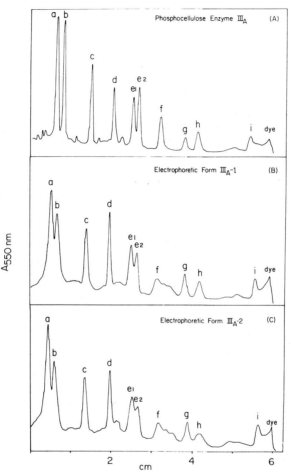

Fig. 5. *Polypeptide compositions of the chromatographic and electrophoretic forms of murine plasmacytoma RNA polymerase III_A. Panel A, chromatographically purified RNA polymerase III_A (phosphocellulose fraction). Panel B, electrophoretic form III_A-1. Panel C, electrophoretic form III_A-2. The electrophoretic forms III_A-1 and III_A-2 were resolved by electrophoresis of chromatographically purified RNA polymerase III_A under non-denaturing conditions as described in Figure 2. Samples of each enzyme were subjected to electrophoresis on 10% polyacrylamide gels in the presence of SDS. After electrophoresis, gels were stained for protein, destained, and scanned at 550 nm.*

in subunit IIIf is also apparent for the phosphocellulose enzyme III_A (Sklar et al., 1975; Sklar and Roeder, 1975; see also below).

These studies suggest that the electrophoretic forms of RNA polymerase III_A (and possibly those of III_B, below) may differ primarily in the charge of a specific subunit(s) or in other minor structural differences not detectable in the present system. Whether these reflect functionally distinct enzyme forms is not yet known but, if so, they could reflect different functional states of an enzyme which transcribes a given class of genes, rather than enzymes which transcribe distinct genes.

As reported previously (Sklar et al., 1975) the chromatographic enzyme forms III_A and III_B have markedly similar subunit compositions. However, when these enzymes are subjected to electrophoresis under denaturing conditions on high resolution polyacrylamide gels a difference can be detected (Fig. 6). Thus, although all other subunits appear identical, the 32,000 dalton subunit ($IIIg_A$) found in form III_A is clearly distinct from the 33,000 dalton subunit ($IIIg_B$) found in form III_B. These observations suggest the possibility of functional differences between RNA polymerases III_A and III_B; for example, in the synthesis of 5S RNA versus tRNAs.

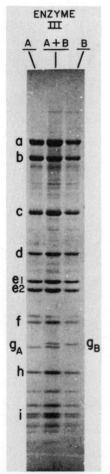

Fig. 6. *Polypeptide compositions of murine plasmacytoma RNA polymerases III_A and III_B. Left enzyme III_A. Right, enzyme III_B. Center, equivalent amounts of enzymes III_A and III_B mixed prior to denaturation. Samples of the chromatographically purified enzymes (phosphocellulose fractions) were subjected to high resolution gel electrophoresis in a 25 cm polyacrylamide slab gel containing a 6 to 11% linear acrylamide gradient in the presence of SDS. For details of the electrophoresis procedures see Sklar and Roeder, 1975.*

Previous comparisons of the class I enzymes and the class II enzymes from different eukaryotes suggested that the structures of the cognate enzymes are in fact conserved to a remarkable extent (reviewed in Chambon, 1975). To establish the generality of these observations for all eukaryotic enzyme classes, it was of interest to examine the structures of the class III enzymes from other eukaryotic cell types, especially in light of the extreme complexity of the mouse enzymes. Single chromatographic forms of RNA polymerase III have been purified from oocytes of *X. laevis* (Roeder, 1974b) and from the posterior silk gland of the silk moth *Bombyx mori* (Sklar, Gage, and Roeder, unpublished observations). A comparison of the polypeptide structures of these enzymes with those of the plasmacytoma enzymes is shown in Figure 7. There is a

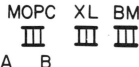

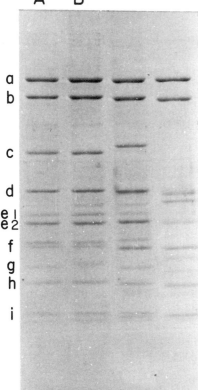

Fig. 7. *Polypeptide compositions of murine plasmacytoma, X. laevis, and B. mori class III RNA polymerases. The mouse plasmacytoma enzymes III_A and III_B were purified as described (Sklar et al., 1975; Sklar and Roeder, 1975). The single chromatographic form of RNA polymerase III present in oocytes of X. laevis (Roeder, 1974b) and the single chromatographic form of RNA polymerase III present in the posterior silk gland of fifth instar B. mori larvae (Sklar, Gage and Roeder, unpublished observations) were purified to apparent chromatographic homogeneity by similar procedures. Samples (3-4 μg) of each of these class III enzymes were subjected to electrophoresis on a 9 cm polyacrylamide slab gel containing a 5 to 10% linear acrylamide gradient in the presence of SDS. Letters indicate the polypeptides previously designated as subunits in the mouse plasmacytoma enzyme (cf. Figure 5). For details of the electrophoresis procedures see Sklar et al., 1975.*

striking similarity between the *X. laevis* oocyte RNA polymerase III and the murine class III enzymes, with only minor differences in molecular weight between some analogous subunits (e.g., the IIIc and the IIIg subunits). Since the amphibian and the murine enzymes are purified by different procedures from vastly different cell types (large nondividing oocytes versus rapidly dividing somatic cells) these findings support the notion that the complex array of similar polypeptides found in each enzyme indeed represent subunits and that they are conserved in evolution. The *B. mori* enzyme also displays a similar polypeptide composition although a polypeptide analogous to subunit IIIc in the murine and amphibian class III enzymes is not evident. The absence of this subunit, as well as the α-amanitin insensitivity of this enzyme (Section III), raises the possibility that certain functional or regulatory features of this enzyme may differ from those of the class III enzymes from higher organisms. However, the major point from these studies is that the structural (and presumably functional) features of the class III enzymes are similar in different organisms, particularly in animal cells.

E. Comparison of Class I, II, and III Structures.

The distinct properties and functions of the class I, II, and III RNA polymerases (Sections III and IV) suggested that these enzymes were structurally distinct. Table III summarizes the molecular weights of the individual polypeptides found in each of the murine class I, II, and III enzymes. Comparison of these data indicates that the large subunits (>100,000 daltons) and some of the low-molecular weight polypeptides appear unique to each enzyme class. In contrast some polypeptides appear on the basis of molecular size to be common to all three enzyme classes, including the 29,000 dalton polypeptide and the 19,000 dalton polypeptide. These conclusions were confirmed in independent studies (Sklar *et al.*, 1975) in which the enzymes were mixed prior to denaturation and electrophoresis to assess more reliably similarities and differences in the sizes of various polypeptides. A comparative analysis (in the same polyacrylamide slab gel) of RNA polymerases I, II, and III$_A$ is also shown in Figure 8. In this particular system both subunits IIIe and IIIf appear heterogeneous (cf. also Figure 6) and it appears that IIIe2 and the lower component in IIIf are similar in size to subunits Id and IId, respectively (see also Sklar *et al.*, 1975). Thus some subunits may be common to only two classes of enzymes. The existence of subunits common to two or three distinct enzyme classes could reflect similar catalytic mechanisms between the different enzymes or the recognition of other common transcription factors. However, more definitive analyses of these polypeptides will be required to establish the hypothesis of common subunits.

The structural studies presented here and elsewhere suggest that the three classes of RNA polymerases are assembled primarily from distinct gene products and that they are not interconvertible by simple structural alterations. Thus, the

TABLE III.
Subunits of the Murine Plasmacytoma Class I, II, and III Polypeptides

	I_A	I_B		II_O	II_A	II_B		III_A	III_B
			II_o	240					
			IIa		205				
Ia	195	195							
			IIb			170			
							IIIa	155	155
			IIc	140	140	140			
							IIIb	138	138
Ib	117	117							
							IIIc	89	89
							IIId	70	70
Ic		61							
							IIIe1	53	53
Id	52	52					IIIe2	49	49
			IId	41	41	41	IIIf	41	41
							$IIIg_B$		33
							$IIIg_A$	32	
Ie	29	29	IIe	29	29	29	IIIh	29	29
			IIf	27	27	27			
			IIg	22	22	22			
If	19	19	IIh	19	19	19	IIIi	19	19
			IIi	16	16	16			

These data summarize the apparent molecular weights of the subunits which are present in the class I, II, and III RNA polymerases and which are indicated in Figures 3-5. RNA polymerase subunits are indicated by Roman numerals (denoting the enzyme class) followed by lower case letters (denoting the specific subunit). The numbers indicate subunit molecular weights x 10^{-3}. Only those polypeptides that remain associated with the individual enzyme forms within each class during electrophoresis under non-denaturing conditions (see text and Figures 2-5) are included here. Enzyme subunits IIh and IIIf each appear to be comprised of at least two separable components (Sklar et al., 1975; Sklar and Roeder, 1975). See text for discussion of those polypeptides which appear common to two or more enzyme classes.

cellular concentrations of the various enzymes may be regulated independently by differential rates of synthesis of class-specific RNA polymerase subunits. Alternatively, the activity or selectivity of the various enzymes may be regulated independently via specific interactions of RNA polymerase subunits with other cellular components. Either of these mechanisms could be involved in regulating the rates of transcription of specific classes of genes.

The class II RNA polymerases are believed to transcribe a much more heterogeneous class of genes than the other RNA polymerases. Hence it is not unexpected that class II RNA polymerases have more complex subunit

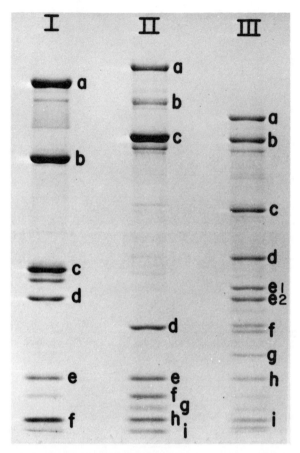

Fig. 8. *Comparison of polypeptide compositions of RNA polymerases I, II, and III. Samples of chromatographic purified enzymes were subjected to electrophoresis in the presence of SDS as described in the legend to Figure 7. The RNA polymerase I sample contained a mixture of the electrophoretic forms I_A and I_B. The RNA polymerase II sample contained a mixture of electrophoretic forms II_A and II_B. The RNA polymerase III sample was enzyme III_A and contained a mixture of electrophoretic forms III_A-1 and III_A-2. The letters denote enzyme subunits whose molecular weights are summarized in Table III. For details of structural comparisons see the text and previous publications (Schwartz and Roeder, 1974; Schwartz and Roeder, 1975; Sklar et al., 1975; Sklar and Roeder, 1975).*

structures than does RNA polymerase I, since the increased complexity offers more potential for interactions with other regulatory components. On the other hand, the class III enzymes appear even more complex than the class II enzymes. Although the significance of these increased structural complexities are not clear, they suggest either that the class III enzymes transcribe heterogeneous gene sequences other than those indicated in Table II, or that there are differences in the modes of transcription and regulation of genes transcribed by RNA polymerases II and III respectively.

V. LEVELS OF CLASS I, II, AND III RNA POLYMERASES DURING ALTERATIONS IN GENE ACTIVITY

Differential alterations in the rates of synthesis of the major classes of RNA (ribosomal RNA, tRNA, HnRNA) have been observed during a number of physiological changes. Since the major RNA species are transcribed by distinct enzymes, it is possible that such modulations in gene activity are effected in part by changes in RNA polymerase concentrations. Alternatively, other factors could modulate the intracellular activities of the respective enzymes. To gain some insight into these questions, this laboratory has investigated RNA polymerase levels in a number of systems in which gross alterations in gene activity have been documented. In these experiments the intracellular RNA polymerases have been solubilized, separated from endogenous templates, resolved chromatographically on DEAE-Sephadex, and the various activities measured in the presence of exogenous DNA or synthetic templates. Under these conditions the activity on non-specific templates appears to be a reflection of RNA polymerase concentration (Schwartz et al., 1974b; Schwartz and Roeder, 1974; Jaehning et al., 1975; Cochet-Meilhac et al., 1974). The following will summarize the results of studies in which the levels of class I, II, and III enzymes have been monitored simultaneously in various systems.

A. RNA polymerase levels during growth transitions.

During the transformation of cultured human lymphocytes by phytohemagglutinin there are gross changes in the rates of synthesis of all major species of RNA as the cells are stimulated to grow and proliferate (for references see Jaehning et al., 1975). Both resting and rapidly proliferating lymphocytes were shown to contain RNA polymerases I, II, and III. However, during four days exposure to phytohemagglutinin the cellular levels of RNA polymerases I and III were found to increase about 17 fold while the level of RNA polymerase II was found to increase about 8 fold (Jaehning et al., 1975), consistent with the increased rates of synthesis of all major classes of RNA. In other studies in various mouse tissues, the levels of RNA polymerases I and III were shown to be considerably higher in tissues with apparently greater rates of cellular growth and proliferation, while the levels of RNA polymerase II were found to be less variable (Schwartz et al., 1974a). These latter observations are consistent with the suggestion that some growth transitions are accompanied by major increases in the rates of synthesis of rRNA, 5S RNA and tRNA with little or no apparent change in the rate of synthesis of HnRNA (Mauck and Green, 1973, 1975). Similarly, altered enzyme levels (especially enzyme I) have been noted in several other cell types in response to growth promoting hormones (Blatti *et al.*, 1970; Sajdel and Jacob, 1971; Smuckler and Tata, 1971; Mainwaring *et al.*, 1971).

B. *RNA polymerase levels during embryonic development.*

A situation which contrasts with the preceding observations is seen during very early embryonic development in *Xenopus laevis* when the absolute and relative rates of synthesis of the major classes of RNA vary enormously (for references see Roeder, 1974b). No RNA synthesis occurs during early cleavage but the genes for HnRNA, tRNA, and rRNA are sequentially activated at subsequent stages, so that by mid gastrula all major classes of genes are active. However, the unfertilized egg contains an enormous cellular excess of RNA polymerases I, II, and III (about 10^4 fold on a per cell basis); and up through the gastrula stage there are no changes in the relative or absolute amounts of these enzymes in the embryo (Roeder, 1974b). The amounts of RNA polymerases I, II, and III per embryo increase only during later embryonic development when, due to continued cell division, the enzyme levels per cell reach those present in adult differentiated cell types.

C. *RNA polymerase levels during lytic virus infection.*

A situation somewhat analogous to that in early amphibian development is seen during the infection of human KB cells by adenovirus 2. As discussed previously (section III) there are qualitative and quantitative alterations in viral gene expression following viral DNA replication. As shown elsewhere, the endogenous RNA polymerase II and III activities in isolated nuclei each increase about 10 fold during the course of viral infection and the products of these activities are predominantly viral mRNA precursors and the small viral RNA species, respectively (Weinmann *et al.*, 1975; Table II). No increase in the endogenous RNA polymerase I activity in isolated nuclei was found, in agreement with the idea that this enzyme does not transcribe viral genes. These endogenous RNA polymerase activities are believed to reflect the intracellular enzyme activities of the enzymes in the intact cell (Weinmann *et al.*, 1975). To investigate the basis for the increased intracellular activities, the cellular levels of the various enzymes were measured with exogenous templates after solubilization. However, in no case (for enzyme I, II, or III) could any change in the cellular enzyme level (concentration) be detected.

D. *Regulation of intracellular RNA polymerase function and specificity.*

The above studies suggest several mechanisms by which the intracellular RNA polymerase activities are regulated. In some instances increased gene activity (i.e. intracellular RNA polymerase activity) can be correlated with increased enzyme levels (concentrations) within the cell, suggesting that specific enzymes may in some instances be limiting and that the level of a specific enzyme may in part regulate the rate of transcription of a specific class of genes. Such changes would appear to affect quantitatively, rather than qualitatively, the transcription of a class of genes by a specific enzyme. However, these changes do not preclude the involvement of other factors in the function or modulation of the activity of a specific RNA polymerase. In the case of RNA polymerase I, for example, an unstable factor has been implicated in the function of this enzyme *in vivo* (Yu and Feigelson, 1972; Gross and Pogo, 1974; Schwartz et al., 1974b; Apriletti and Penhoet, 1975). Although factors which alter the activity of RNA polymerase II *in vitro* have been isolated (cf. Stein and Hausen, 1970; Sugden and Keller, 1973) their *in vivo* functions(s) remain to be elucidated.

In other instances (e.g. embryonic development, lytic virus infection) qualitative and quantitative changes in gene activity, and hence in intracellular RNA polymerase activities, are not associated with noticeable differences in RNA polymerase levels. In these cases other factors must clearly modulate the intracellular activity and specificity (site selectivity) of the various RNA polymerases. Furthermore one expects *a priori* that such regulatory or specificity factors must exist since alterations in enzyme levels *per se* cannot effect selective gene activation (differential gene transcription) within a heterogeneous class of genes which are transcribed by a common RNA polymerase. This is especially significant in the case of the class II enzymes which effect the differential transcription of a very heterodisperse class of genes. Similarly, differential transcription of the 5S RNA and individual tRNA genes by the class III enzymes would necessitate other specificity factors for regulation of this class of enzymes. Whether additional specificity elements are necessary for the class I enzymes is unknown since it is uncertain whether genes other than those coding for rRNA are transcribed by these enzymes and if so whether they are subject to differential transcription. As discussed earlier (Introduction, Section I), these additional specificity factors may be involved with chromatin structural modifications or they may influence RNA polymerase-DNA interactions more directly by interaction with either or both of these components.

VI. CONCLUSIONS

Eukaryotic cells contain multiple molecular species of nuclear RNA polymerase which fall into three major classes (I, II, and III). The class I, II, and III enzymes have distinct catalytic properties, markedly distinct subunit structures, and specific functions in the synthesis of the major classes of RNA. In constrast, the individual enzyme forms within a given class have similar catalytic properties and in each case differ in only a single subunit, suggesting that they have related, but nonetheless distinct, functions.

During some physiological transitions, alterations in gene activity are accompanied by changes in enzyme levels, suggesting that quantitative changes in the transcription of specific classes of genes may be effected by changes in enzyme levels. However, during other physiological transitions, quantitative and qualitative changes in gene activity are not accompanied by changes in enzyme levels. These latter observations plus the existence of only a limited number of cellular RNA polymerases suggest that additional factors must also regulate the activity and/or selectivity of the various RNA polymerases.

In considering the possible mechanisms involved in transcriptional regulation it is relevant to consider the structural complexity of the eukaryotic RNA polymerases. Each class I, class II or class III enzyme contains approximately six, eight, or ten polypeptides, respectively. Since the complexity of these enzymes is considerably greater than that necessary for the recognition of a specific site on the DNA by an RNA polymerase (Chamberlin and McGrath, 1970) it seems likely that the eukaryotic enzymes play more than a passive role in the regulation of transcription. Eukaryotic enzymes might require more polypeptides for mechanical aspects of transcription than do bacteriophage (Chamberlin and McGrath, 1970) or bacterial (Burgess, 1971) enzymes, especially if active genes are associated with a chromatin subunit structure (Axel, Cedar, and Felsenfeld, 1975). However, the polypeptides involved in such catalytic functions might be expected to be common to the class I, II, and III RNA polymerases and apparently only two polypeptides fall into this category. Thus, one may speculate that many of the polypeptides found in the eukaryotic RNA polymerases are in fact involved in regulatory interactions of the enzymes with specific cellular components, and that these interactions serve to modulate the activity or the site selectivity of the enzymes. Such interactions might occur concomitantly with, and represent an essential feature of, chromatin structural modifications or they might occur at a subsequent stage, effecting a finer level of control over that imposed by the chromatin structure alone (see Introduction).

Finally, it is hoped that the present investigations and understanding of the general properties, functions, structures, and changes in the cellular levels of the respective class I, II, and III RNA polymerases will facilitate the detection and characterization of undefined cellular components which may be involved in the activation and regulation of specific genes.

ACKNOWLEDGEMENTS

This work was supported by research grants from the National Institutes of Health (CA 16640 and GM 19096) and from the National Science Foundation (BMS 74-24657). R.G.R. is Research Career Development Awardee (GM-70661), L.B.S. a Medical Scientist Trainee (GM-02016), and V.E.F.S. a Predoctoral Trainee (GM-1311) of the NIH.

REFERENCES

Adman, R., Schultz, L. D. and Hall, B. D. (1972). *Proc. Nat. Acad. Sci. U.S.A.* **69**, 1702–1706.
Axel, R., Cedar, H. and Felsenfeld, G. (1973). *Proc. Nat. Acad. Sci. U.S.A.* **70**, 2029–2032.
Axel, R., Cedar, H. and Felsenfeld, G. (1975). *Biochemistry* **14**, 2489–2495.
Apriletti, J. W. and Penhoet, E. E. (1974). *Virology* **61**, 597–601.
Blatti, S. P., Ingles, C. J., Lindell, T. J., Morris, P. W., Weaver, R. F., Weinberg, F. and Rutter, W. J. (1970). *Cold Spr. Harbor Symp.* **35**, 649–657.
Bonner, J. E., Dahmus, M. E., Fambrough, D., Huang, R. C. C., Marushige, K., and Tuan, D. H. Y. (1968). *Science* **159**, 47–56.
Brown, D. D., Wensink, P. C. and Jordan, E. (1971). *Proc. Nat. Acad. Sci. U.S.A.* **68**, 3175–3179.
Burgess, R. (1971). *Ann. Rev. Biochem.* **40**, 711–740.
Chamberlain, M. and McGrath, J. (1970). *Cold Spr. Harbor Symp.* **35**, 259–262.
Chambon, P. (1975). *Ann. Rev. Biochem.* **44**, 613–638.
Chesterton, C. J. and Butterworth, P. H. W. (1971). *Eur. J. Biochem.* **19**, 232–241.
Clarkson, S. G. and Birnsteil, M. L. (1973). *Cold Spr. Harb. Symp.* **38**, 451–459.
Cochet-Meilhac, M., Nuret, P., Courvalin, J. C. and Chambon, P. (1974). *Biochem. Biophys. Acta* **353**, 185–192.
Darnell, J. E., Jelinek, W. R. and Molloy, G. R. (1973). *Science* **181**, 1215–1221.
Gilmour, R. S. and Paul, J. (1973). *Proc. Nat. Acad. Sci. U.S.A.* **70**, 3440–3442.
Gissinger, F. and Chambon, P. (1972). *Eur. J. Biochem.* **28**, 277–282.
Green, M. (1970). *Ann. Rev. Biochem.* **39**, 735–755.
Gross, K. J. and Pogo, A. O. (1974). *J. Biol. Chem.* **249**, 568–576.
Jaehning, J. A., Stewart, C. C. and Roeder, R. G. (1975). *Cell* **4**, 51–57.
Kedinger, C. and Chambon, P. (1972). *Eur. J. Biochem.* **28**, 283–290.
Kedinger, C., Gissinger, F. and Chambon, P. (1974). *Eur. J. Biochem.* **44**, 421–436.
Lindell, T. J., Weinberg, F., Morris, P. W., Roeder, R. G. and Rutter, W. J. (1970). *Science* **170**, 447–449.
Mainwaring, W. I. F., Mangan, F. R. and Peterken, B. M. (1971). *Biochem. J.* **123**, 619–628.
Marzluff, W. F., White, E. L., Benjamin, R. and Huang, R. C. (1975). *Biochemistry* **14**, 3715–3724.
Mauck, J. C. and Green, H. (1973). *Proc. Nat. Acad. Sci. U.S.A.* **70**, 2819–2822.
Mauck, J. C. and Green, H. (1974). *Cell* **3**, 171–177.
Ohe, K. and Weissman, S. M. (1971). *J. Biol. Chem.* **246**, 6991–7009.
Phillips, J. and Forrest, H. S. (1973). *J. Biol. Chem.* **248**, 265–269.
Price, R. and Penman, S. (1972a). *J. Virology* **9**, 621–626.
Price, R. and Penman, S. (1972b). *J. Mol. Biol.* **70**, 435–470.
Reeder, R. H. and Roeder, R. G. (1972). *J. Mol. Biol.* **67**, 433–441.
Roeder, R. G. (1969). Multiple RNA polymerases and RNA synthesis in eukaryotic systems, Ph.D. thesis, University of Washington.
Roeder, R. G. (1974a). *J. Biol. Chem.* **248**, 241–248.
Roeder, R. G. (1974b). *J. Biol. Chem.* **248**, 249–256.
Roeder, R. G. and Rutter, W. J. (1969). *Nature* **224**, 234–237.
Roeder, R. G. and Rutter, W. J. (1970a). *Proc. Nat. Acad. Sci. U.S.A.* **65**, 675–682.
Roeder, R. G. and Rutter, W. J. (1970b). *Biochemistry* **9**, 2543–2553.
Sajdel, E. M. and Jacob, S. T. (1971). *Biochem. Biophys. Res. Commun.* **45**, 707–715.

Schwartz, L. B. and Roeder, R. G. (1974). *J. Biol. Chem.* **249**, 5898–5906.
Schwartz, L. B. and Roeder, R. G. (1975). *J. Biol. Chem.* **250**, 3221–3228.
Schwartz, L. B., Sklar, V. E. F., Jaehning, J., Weinmann, R. and Roeder, R. G. (1974a). *J. Biol. Chem.* **249**, 5889–5897.
Schwartz, L. B., Lawrence, C., Thach, R. E. and Roeder, R. G. (1974b). *J. Virology* **14**, 611–619.
Seifart, K. H., Benecke, B. J. and Juhasz, P. P. (1972). *Arch. Biochem. Biophys.* **151**, 519–532.
Seifart, K. H. and Benecke, B. J. (1975). *Eur. J. Biochem.* **53**, 293–300.
Sergeant, A., D'Halluin, J. C., Verbert, A. P. and Krsmanovic, V. (1974). *Arch. fur die gesamte Virusforschung* **45**, 294–297.
Sharp, P. A., Gallimore, P. H. and Flint, S. J. (1974). *Cold Spr. Harb. Symp.* **39**, 457–474.
Sklar, V. E. F. and Roeder, R. G. (Submitted for publication).
Sklar, V. E. F., Schwartz, L. B. and Roeder, R. G. (1975). *Proc. Nat. Acad. Sci. U.S.A.* **72**, 348–352.
Smuckler, E. A. and Tata, J. R. (1971). *Nature (London)* **234**, 37–39.
Steggles, A. W., Wilson, G. N., Kantor, J. A., Picciano, D. J., Falvey, A. K. and Anderson, W. F. (1974). *Proc. Nat. Acad. Sci. U.S.A.* **71**, 1219–1223.
Stein, H. and Hausen, P. (1970). *Cold Spr. Harb. Symp. Quant. Biol.* **35**, 709–717.
Sugden, B. and Keller, W. (1973). *J. Biol. Chem.* **248**, 3777–3788.
Wallace, R. D. and Kates, J. (1972). *J. Virol.* **9**, 627–635.
Weaver, R. F., Blatti, S. P. and Rutter, W. J. (1971). *Proc. Nat. Acad. Sci.* **U.S.A. 68**, 2994–2999.
Weinberg, R. A. and Penman, S. (1968). *J. Mol. Biol.* **38**, 289–304.
Weinmann, R. and Roeder, R. G. (1974). *Proc. Nat. Acad. Sci. U.S.A.* **71**, 1790–1794.
Weinmann, R., Raskas, H. and Roeder, R. G. (1974). *Proc. Nat. Acad. Sci. U.S.A.* **71**, 3426–3430.
Weinmann, R., Jaehning, J. A., Raskas, H. J. and Roeder, R. G. (1975a). (Submitted for publication).
Weinmann, R., Brendler, T., Raskas, H. J. and Roeder, R. G. (1975b). (Submitted for publication).
Yu, F. L. and Feigelson, P. (1972). *Proc. Nat. Acad. Sci. U.S.A.* **69**, 2833–2837.
Zylber, E. and Penman, S. (1971). *Proc. Nat. Acad. Sci. U.S.A.* **68**, 2861–2865.

BIOCHEMICAL EVENTS IN mRNA FORMATION IN MAMMALIAN CELLS

James E. Darnell, Warren Jelinek, Larry Puckett,[2]
Eva Derman[3] and Steven Bachenheimer

Rockefeller University
New York, N.Y. 10021

I. INTRODUCTION

Central to the interpretation of embryological (developmental) events in molecular terms is an understanding of how the *supply* of mRNA is regulated. Conventional thinking about the supply of mRNA has been molded by the discovery that in bacteria and phage-infected bacteria, proteins interact with DNA to either promote or restrict transcription (Jacob and Monod, 1961; Zubay et al., 1970; Greenblatt and Schleif, 1971). Because a sensitive regulation of mRNA formation is coupled with a relatively rapid turnover of existing mRNA molecules (Levinthal et al., 1963), the *supply* of mRNA is essentially governed at the level of synthesis. Implicit in the bacterial model is that every mRNA region which is transcribed is also translated.

In mammalian cells, a number of post-transcriptional RNA modifications occur in the formation of mRNA prior to its participation in protein synthesis (Darnell et al., 1973). Since these biochemical modifications also occur in HnRNA[1] molecules, a precursor role for the RNA involving specific cleavage to yield mRNA has been proposed. With the interposition of several biochemical events between the transcription and eventual function of mRNA, the potential arises for *post-transcriptional regulation* in mRNA supply. If each of several steps is necessary for mRNA formation, and cells have the option to decide yes or no at each step, post-transcriptional regulation in the delivery of already transcribed mRNA sequences is possible.

Also a third level of regulation almost certainly occurs in mammalian cells-i.e., – *translational regulation*. At this level decisions might be made involving either the number of successful ribosomal engagements by a given mRNA and/or the lifetime of the mRNA.

Many laboratories, including our own, have for the past several years been studying the biochemistry of post-transcriptional events in mRNA formation in mammalian cells with the hope of eventually determining, what, if any, role these steps play in regulation of mRNA supply. Considerable progress has been made in outlining necessary steps in mRNA formation; little is yet known about whether these steps are points of regulation.

Our central concern has been the origin of mRNA — Does HnRNA, the primary gene product, after post-transcriptional modification, give rise to mRNA?

It is appropriate first to review what type of evidence is necessary to prove such a relationship and why incontrovertible evidence of the HnRNA → mRNA pathway is exceedingly difficult to achieve. Then the evidence at hand that some HnRNA molecules do represent mRNA precursors will be given.

Two types of evidence are required to prove that a larger RNA molecule is a precursor to a shorter RNA product and a third type of evidence is needed to understand the reaction in detail. First, the identical primary nucleotide sequence found in the product must be present, in company, of course, with additional covalently linked sequences, in the precursor molecule. Second it must be shown that the only form in which the eventual product is actually synthesized is as the larger precursor, followed by processing, including cleavage. That is, the mere existence of possible precursor molecules, together with shorter products, does not constitute sufficient evidence that the larger molecules are, in fact, precursors. Basically what is demanded for this second type of evidence is radioactive labeling of the precursor prior to labeling of final product, followed if possible by demonstration of a transfer of label to the shorter form in the absence of further synthesis. The first RNA molecule shown to function as a precursor was r-pre-RNA from HeLa cells (Scherrer and Darnell, 1962; Scherrer et al., 1963; Perry, 1962). Chemical relatedness involving methylated oligonucleotide overlap between 45S r-pre-RNA and 28 and 18S rRNA has been demonstrated (Wagner et al., 1967; Maden, 1971), in addition to the original kinetic labeling studies that showed the 45S was synthesized first followed by a transfer of radioactivity to 28 and 18S (Scherrer et al., 1963). Similar but less extensive evidence has also established a class of tRNA precursors, pre-tRNA, in cultured cells (Bernhardt and Darnell, 1968; Burdon et al., 1967).

A final type of evidence, very consoling to devout biochemists and obviously important to a final understanding of precursor conversion, but *not* strictly necessary for the proof of a precursor role of larger molecules, is the demonstration of cleavage into a finished product of purified precursor by purified enzymes. Such specific enzymatic cleavage in the conversion of precursors into final products has recently been achieved with bacterial and bacteriophage pre-t-RNA to tRNA. and T-7 specific RNA into T-7 mRNA (reviewed by Robertson and Dickson, 1974). Cell-free cleavage of mammalian pre-tRNA with crude cell extracts was, in fact, the first cell-free RNA cleavage to be reported although the precision of this cleavage was not possible to assess (Mowshowitz, 1969). In view of all these findings, the proposal of several years ago that the HnRNA might be cleaved to yield mRNA is not by this time without precedent (Scherrer et al., 1963; Penman et al., 1963).

II. EVIDENCE FOR SEQUENCES SHARED BY HnRNA AND mRNA

A. Sequences Found in the Coding Portion of mRNA.

Several years of attempts to show sequence similarity or difference of mRNA and HnRNA by hybridization to total cell DNA produced only evidence that both species contained regions which rapidly hybridized (i.e., sequences drawn from repetitive DNA sites; Britten and Kohne, 1968) but mostly regions which slowly hybridized (i.e., sequences from less repeated or unique DNA sequences). No conclusion about shared sequences was possible (Shearer and McCarthy, 1967, 1970; Whiteley, McCarthy and Whiteley, 1966; Soeiro and Darnell, 1970; Pagoulatos and Darnell, 1970; Scherrer et al., 1970).

The first success in relating HnRNA to mRNA by criteria other than base compositional similarity came through examination of the HnRNA of cells transformed by the small DNA viruses (Sambrook et al., 1968); virus-specific RNA was indeed found in dimethylsulfoxide (Me_2SO)-stable high molecular weight HnRNA as well as in lower molecular weight virus-specific mRNA of discrete size(s) (Lindberg and Darnell, 1970; Tonegawa et al., 1970; Wall et al., 1973). It was also demonstrated that cells infected with viruses whose DNA was transcribed in the cell nucleus contained high molecular weight nuclear RNA while cytoplasmic viral mRNA was smaller in size (Roizman et al., 1970; Parsons et al., 1971; Wall et al., 1972).

Additional recent evidence has shown that HnRNA from birds and mammals contain hemoglobin mRNA sequences as detected both by molecular hybridization (Melli and Pemberton, 1972; Imaizumi et al., 1973; Lanyon et al., 1975) and translation into hemoglobin (Ruiz-Carillo et al., 1973). One important point about the content of either virus-specific or hemoglobin sequences in HnRNA has recently emerged. When the total amount of virus-specific sequences was measured either by hybridization to virus DNA of steady-state labeled nuclear RNA from transformed cells (Georgieff et al., 1974 and Fig. 1) or by measuring the amounts of radioactive hemoglobin cDNA (made as a complementary copy of hemoglobin mRNA by reverse transcriptase; Verna et al., 1973; Ross et al., 1973; and Kacian et al., 1973) bound by nuclear RNA, very little (10% or less) of the nuclear sequences were larger than 10,000 nucleotides. A majority of the virus-specific sequences in nuclear RNA from transformed cells sedimented similarly to the cytoplasmic sequences. In the case of RNA complementary to hemoglobin cDNA various

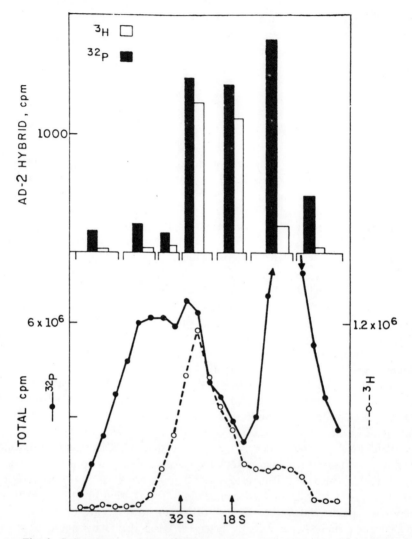

Fig. 1. *Sedimentation of total HnRNA and "small" HnRNA in dimethylsulfoxide (Me_2SO). Total ^{32}P HnRNA and a sample of 15-32S ^3H-labeled HnRNA from adenovirus transformed cells were sedimented through (Me_2SO) and total radioactivity and Ad2-specific RNA measured.*

reports have demonstrated only a small amount (~5%) of the hemoglobin "HnRNA" larger than 30S, (Imaizumi et al., 1973; Lanyon et al., 1975) but a substantial amount sedimenting at about 12-15S (McNaughton et al., 1974) i.e., about three times larger than the finished product but almost none of the finished "9S" product in the nucleus.

Clearly these studies (together with the aforementioned translation studies) establish that sequences reponsible for virus protein and hemoglobin synthesis can be found in larger molecules; they do not in any way prove that the larger molecules are precursor to the smaller cytoplasmic mRNA. Furthermore, because the fraction of the total nuclear RNA represented by any single RNA type is so small (10^{-4} to 10^{-5} of the total), very brief labels are not practical to determine if the large molecule is labeled first.

B. *Post-Transcriptional Additions Shared by HnRNA and mRNA.*

1. *Poly (A).* By 1971 (Edmonds et al., 1971; Kates, 1970; Lim and Canellakis, 1970; Lee et al., 1971; Darnell et al., 1971a) it was realized that some HnRNA and most mRNA molecules shared an unusual polynucleotide segment — a 3' post-transcriptionally added, ~200 nucleotide segment of polyadenylic acid (Molloy and Darnell, 1973; Molloy et al., 1972; Sheldon et al., 1972; Mendecki et al., 1972). We concluded that this material was synthesized in the nucleus by 3' addition to HnRNA (Darnell et al., 1971b; Jelinek et al., 1973) and that failure of this addition prevented mRNA from appearing in the polyribosomes (Darnell et al., 1971; Adesnik et al., 1972).

Most of the newly formed poly(A) is contained in HnRNA larger than mRNA (Jelinek et al., 1973; Derman and Darnell, 1974 and Fig. 2). A sub-

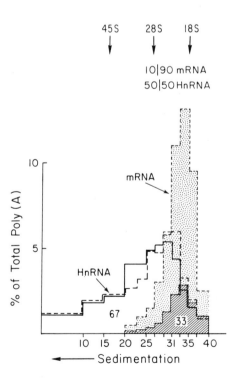

Fig. 2. *Poly(A) content of Me_2SO stable HnRNA and mRNA.*

Cells which had been treated with actinomycin (0.04μg/ml) to suppress rRNA synthesis were labeled with [³H]adenosine either for 100 seconds or 90 minutes. Nuclear RNA was extracted from both cultures and cytoplasmic RNA from the culture labeled for 90 minutes. The RNA was dissolved in 90% Me_2SO, dimethylformamide and 5% NETS buffer (.01 M EDTA, 0.01 M Tris, pH 7.4; and 0.2% SDS), diluted with 2 volumes of 0.05 NETS buffer (same as ETS buffer plus 0.05 M NaCl) and sedimented through identical sucrose gradients (15-30% sucrose in 0.05 NETS). Poly(A) content was then analyzed and the profiles of percent of total poly(A) are plotted for the long-labeled (—) and short-labeled (———) nuclear samples as well as the long labeled cytoplasmic sample. The cytoplasmic sample is plotted twice; the lower scale equalizes the peak height of the cytoplasmic maximum with the nuclear distribution to show how much of the nuclear poly(A) (67%) is contained in molecules larger than the cytoplasmic distribution.

stantial fraction of newly formed poly(A) appeared in mRNA during a "chase" with 3'-deoxyadenosine, a drug which prevented further poly(A) synthesis (Jelinek et al., 1973) and during cell growth the accumulation of labeled cytoplasmic poly(A) and total label in mRNA indicated a much greater conservation of poly(A) than of total HnRNA (Jelinek et al., 1973). Thus, the apparent conservation of this unusual polynucleotide region shared by mRNA and HnRNA seemed to offer the best proof that HnRNA does in fact give rise to mRNA.

Controversey over this interpretation arose when Perry, Kelley and Latorre (1974) examined the kinetics of appearance of poly(A) in the nucleus and in the cytoplasm of L cells and tried to fit the data to equations describing a simple precursor-product relationship, knowing the steady state amounts of nuclear and cytoplasmic poly(A) and using a recently obtained estimate of mRNA half-life. The equations assumed a single entry time of nuclear poly(A) containing material into the cytoplasm and a half-life of mRNA of 6 to 10 hours. Within the restraints of the described equations, it appeared to Perry and his coworkers that a) only a small amount of nuclear poly(A) could contribute to the cytoplasm while the remainder must turnover and b) cytoplasmic poly(A) synthesis was proposed because of the rapid appearance of cytoplasmic poly(A) prior to maximal nuclear poly(A) labeling.

Reconsideration of the assumptions made by Perry and his coworkers in their kinetic analysis led us to look for short-lived mRNA. Clearly, if short-lived RNA exists then a more rapid appearance of cytoplasmic poly(A) would be expected. In addition, if some mRNA molecules spent only a short time in the nucleus after poly(A) addition while others spent a longer time then a rapid cytoplasmic poly(A) increase could also be explained. Short-lived mRNA was, in fact, observed (Puckett et al., 1975 and Fig. 3) and a model was described in which poly(A) synthesis takes place only as a nuclear precursor to cytoplasmic poly(A) contained in three average mRNA classes, 40% with a half-life of 16 hours, 40% with a half-life of 7 hours, and 20% with a half-life of 1 hour. The actual accumulation of nuclear and cytoplasmic poly(A) fit the curves predicted by this model quite well (Fig. 4). We cannot conclude however, as a proved fact that all nuclear poly(A) does exit to the cytoplasm, only that it is possible. Since about 70-80% of the nuclear poly(A) is in molecules larger than mRNA, cytoplasmic appearance of the majority of poly(A) would indicate the necessity of HnRNA cleavage to yield mRNA. In the worst case analysis, however, this leaves open the possibility that if only 20-30% of the nuclear poly(A) exited to the cytoplasm, HnRNA might not be an mRNA precursor. Further evidence on the relationship of HnRNA to mRNA was needed therefore.

However, before going to a discussion of another type of biochemical marker shared by HnRNA and mRNA, we have also reviewed the suggestions

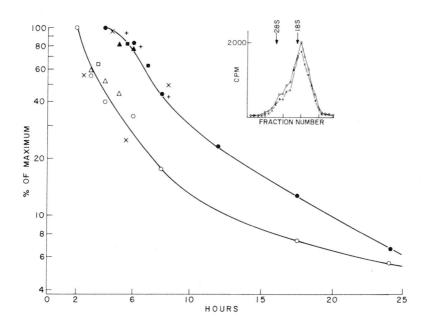

Fig. 3. *Turnover of [³H] guanosine-labeled mRNA. Cells pulse-labeled with [³H] guanosine and chased in unlabeled medium were assayed for radioactivity in cytoplasmic poly(A)-terminated molecules at indicated time intervals. Different symbols represent different experiments. Insert shows zonal sedimentation profile of poly(U)-sepharose bound fraction and lack of rRNA contamination (0, 2 hr; x, 5 hr samples, same number of cpm was placed on the gradient for analysis). Open symbols, cpm in mRNA compared to cpm in 18S rRNA, with the 2 hr sample considered as 100%. Closed symbols, cpm in mRNA 4×10^7 cells, with the 4-hr sample considered as 100%.*

that some cytoplasmic poly(A) synthesis does occur in somatic cells. Our earlier conclusion that all or almost all the 200 nucleotide, 3' OH, poly(A) segment originated in the nucleus rested on finding over 90-95% of the labeled poly(A) attached to HnRNA when cells were labeled for 1.5 min or less (Darnell et al., 1971; Jelinek et al., 1973). Unless there were a special cytoplasmic ATP pool, significant amounts of cytoplasmic poly(A) synthesis would have been observed in the short-label experiment. Figure 5 diagrams a short-label experiment where the predominant labeled poly(A) is nuclear and also

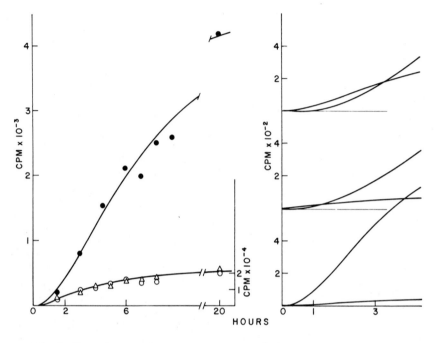

Fig. 4. *The real and predicted accumulation of [^{3}H] adenine in the acid-soluble pool and in nuclear and cytoplasmic poly(A). Left Panel: Cells growing in [^{3}H] adenine accumulated radioactivity in the acid-soluble pool (△, right axis) and in nuclear and cytoplasmic poly(A) (left axis: ●, cytoplasmic; ○ nuclear; cpm 10^7 cells is presented) as shown. Right panel: Experimental data for the accumulation in HeLa cells of radioactivity into the acid-soluble pool were used to calculate the hypothetical curves of accumulation into nuclear and cytoplasmic poly(A), assuming there were three classes of mRNA with turnover times of 1, 7, and 24 hr. In all cases the cytoplasmic curve begins low and ends as the highest curve. Top, 24 hr half-life; middle, 7 hr half-life; and bottom, 1 hr half-life. The sum of the three curves is the solid line in left panel (See Puckett et al., 1975).*

where a faster migrating cytoplasmic adenine-rich species is also evident. This fraction is the dominant cytoplasmic fraction after the 1.5 min label. Poly(A) of this size has been found in mitochondria (Perlman et al., 1973), suggesting the experiment shown in Figure 6. The cytoplasm from cells labeled for 3 min was fractionated into a pellet containing mitochondria plus other membrane associated structures and "free" cytoplasmic structures. Poly(A) was analyzed in each sample and about 50% of the total radioactive cytoplasmic poly(A) can be seen to be in the 50 nucleotide mitochondrial poly(A), whereas the steady state amount of this material is about 15-20% of the total labeled cytoplasmic poly(A). In this experiment, about 3 times as much of the 200-nucleotide poly(A) was still in the nucleus as in the cytoplasm. Since mito-

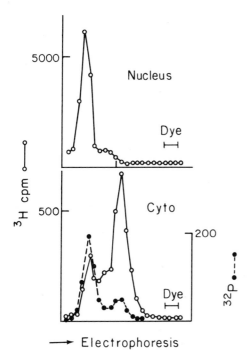

Fig. 5. *Cellular distribution on poly(A): 1.5 minute label.*

Cells were labeled with $[^3H]$-adenosine for 1.5 min and nuclear and cytoplasmic RNA prepared and digested with T1 RNase (Molloy et al., 1974). Poly(A) was selected by poly(U) sepharose chromatography and subjected to gel electrophoresis. ^{32}P labeled cytoplasmic RNA was added to the ^3H cytoplasmic sample to mark the position of cytoplasmic poly(A) (peak in Fraction 6) and mitochondrial poly(A) (peak in Fraction 15).

chondrial poly(A) can become so highly labeled in a very brief pulse while the nucleus contains the great majority of the 200 nucleotide poly(A), any direct cytoplasmic poly A synthesis would require a third non-mixing pool. Because 95% of the newly formed poly(A) is nuclear, (Fig. 5) we again conclude that the 200 nucleotide poly(A) found in mRNA originates in the nucleus.

In addition, the gel electrophoresis profile of poly(A) in the cytoplasm after a 3 minute label showed over 90% of the labeled non-mitochondrial poly(A) segments to be larger than ~150 nucleotides. It has been suggested that cytoplasmic poly(A) "repair" or end-addition goes on at a rate comparable to the rate of poly(A) synthesis (Diez and Brawerman, 1974). The number average distribution of poly(A) in mRNA at steady state vastly favors short poly(A) segments (Sheiness and Darnell, 1972; Sheiness et al., 1975) and if "repair" occurs on all poly(A) segments, a brief label should show heavier incorporation into short poly(A) segments. The data in Figure 5 fails to show evidence of any such distribution of short-labeled poly(A) segments. Any repair synthesis (and some may exist) thus proceeds at a rate considerably slower than the rate of renewal of cytoplasmic poly(A) due to cell growth and mRNA turnover.

We now return to a further consideration of shared chemical features of HnRNA and mRNA.

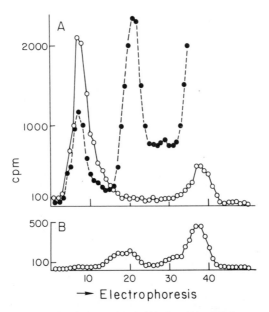

Fig. 6. *Cytoplasmic distribution of poly(A): 3 minute label.*

Cells labeled with [³H]adenosine for 3 min were broken by hypotonic swelling and fractionated into mitochondrial and free cytoplasmic portions. RNA was extracted from each and the total poly(A) containing RNA selected by poly(U) sepharose chromatography from the mitochondiral fraction. The free cytoplasmic poly(A) containing RNA was eluted from poly(U) sepharose in two samples (A) RNA eluted between 20-50% formamide in ETS buffer; (B) RNA eluted between 10-20% formamide in ETS buffer).

Poly(A) was then assayed by gel electrophoresis after digestion of each sample (–•–) "mitochondrial" fraction; (–○–) cytoplasmic fractions.

2. *Methyl groups in mRNA and HnRNA.* The addition of methyl groups to bases or to the ribose moieties of nucleotides is one of the most widespread post-transcriptional modifications of RNA molecules. Perhaps because of early findings that methylation inhibited translation, or because certain virus RNAs that could serve as mRNA were said to lack methyl groups, or because HnRNA was found to contain many fewer groups than r-pre-RNA (Wagner et al., 1967), it was widely assumed that mRNA lacked methyl groups. In spite of previous beliefs, when methods for examining purified cellular mRNA became available (Nakazato et al., 1973), Perry and Kelley (1974) tested for the incorporation of label into mRNA from methyl-labeled methionine, the most effective methyl donor for r-pre-RNA and pre-t-RNA. They were rewarded by discovering incorporation consistent with several methyl groups per molecule. They then examined HnRNA and found some evidence for methyl-labeling there as well. Shortly thereafter Shatkin, Furuichi and colleagues (1975a and b), and Wei and Moss (1975) also found reovirus,

cytoplasmic polyhedrosis virus and vaccinia virus mRNA molecules contained methyl groups at their 5' termini.

Rottman, Shatkin and Perry (1974) suggested that perhaps methyl group addition would prove an aid in studying cellular mRNA formation and that, in fact, if cell mRNA methylation was 5' terminal, perhaps methylation of HnRNA would function as a cleavage (processing) signal to create a new 5' end of mRNA.

Several groups, including our own, in collaboration with Furuichi, Shatkin and colleagues (1975c) have now identified the nature and location of methyl groups in cellular mRNA. The 5' terminus of HeLa cell mRNA has the general structure $m^7G(5')pppN^mpNp$ or $pppNmpNp$ or $m^7G(5')pppN^mpN^mpNp$ — where N can be any nucleotide. This blocked terminal group plus the neighboring 2'0-ribose methylation of the first or first and second nucleotides of the mRNA chain render the terminal 3 or 4 nucleotides resistant to alkali or usual endonucleolytic cleavage. This resistant structure has been referred to as a "cap". In order to quantitate the number of caps per mRNA molecule, ^{32}P mRNA selected by poly(U) sepharose was analyzed for total radioactivity and radioactivity in "caps" and poly(A) (Fig. 7). There were 5-6 CPM in caps compared to 1500 CPM in the total chain and 150 CPM in poly(A). Since the number average for the length of HeLa cell mRNA appears to be about 1500 by sucrose gradient sedimentation analysis and the

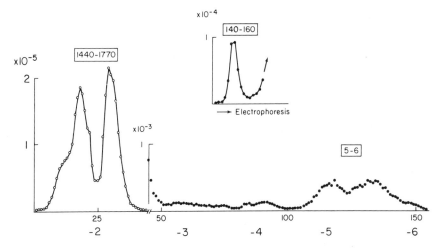

Fig. 7. *"Cap" content of HeLa cell mRNA.*
[^{32}P] mRNA prepared from cells labeled for 5 hrs was digested with pancreatic RNase followed by T2 RNase (Furuichi, et al., 1975a) and subjected to DEAE column chromatography. A small portion of the mRNA was analyzed for poly(A) content. Assuming the cap (−5 to −6) structure to contain 5−6 phosphates the ratio of phosphates in poly(A) and total chains were calculated and appear in insets.

number average length of poly(A) after labeling cells for 5 hours is about 150, it appears that every poly(A) containing mRNA molecule has a cap.

In addition methyl-labeled HeLa cell mRNA contained about 30% of the total methyl counts in N-6 mAp with the remaining 70% in caps. If every mRNA molecule contains one cap (with two or three methyl groups) then it appears that most or all mRNA molecules also contain one N-6 mAp residue. The presence of N-6 mAp as an isolated base-methylated component in cellular RNA takes on added significance because this component is absent from reovirus, CPV, vaccinia and VSV mRNA molecules (Furuichi et al., 1975a, b; Wei and Moss, 1975). All of these viruses replicate in the cytoplasm and their mRNA molecules are made by virion polymerases and directly capped without a reduction in size. In contrast, SV40 (Lavi and Shatkin, 1975) and adenovirus mRNA molecules (Sommers et al., 1975) are transcribed from viral DNA in the cell nucleus and are probably processed from high molecular weight RNA. Both these virus mRNA's contain N-6 methyl adenylic acid and in adenovirus mRNA the ratio is equimolar with caps.

In order to determine the location of N-6 methyl adenylic acid within the mRNA molecule, HeLa cell mRNA was randomly fragmented to a size of about 500 nucleotides and the 3' portion reselected on poly(U) sepharose. The bound fraction, i.e., the poly(A) containing 3' ends as well as the fragments including the 5' portions of the molecules were then analyzed for methyl groups (Furuichi et al., 1975c). Only 5-8% of the labeled methyl groups were in the bound fractions (Table 1) and methyl-label was present in

TABLE I.
Distribution of Methyl ^3H-Labeled Nucleotide in Poly(U)-Bound and Unbound Fragments of T_1 Digested mRNA.

	^{32}P		^3H-Methyl		Ratio mononucleotides cap
	Bound	Unbound	Bound	Unbound	
Exp. 1	80,000* (29)+	200,000 (71)	1000 (6)	16,000 (94)	1:2
Exp. 2	88,000 (32)	184,000 (68)	1250 (4)	35,000 (96)	1:2

*CPM
+Percent of total CPM

^{32}P and [methyl-^3H] methionine-labeled mRNA (less than 50μg total) was mixed with 200 μg of yeast RNA precipitated with ethanol and redissolved in 1.9 ml steterile 1μM tris, pH 7, 1 μM EDTA and 0.0075 ml of dilute T_1 RNase solution (enzyme was dissolved at 100 units/ml in 0.05 M acetate, pH 5, heated at 80°C - 10 min; this stock solution was diluted 100-fold for this digestion) was added for 5 min at 37°C. The RNA was phenol extracted and excess phenol removed by several chloroform extractions. Poly(U)-sepharose selection was carried out, and the unbound fraction was analyzed by DEAE-cellulose chromatography.

both caps and N-6 mAp in a 2/1 ratio in the fragments containing the 5' portions of the mRNA. Since methyl groups are probably not in the coding region and are not in the poly(A) containing region of the mRNA, the N-6 mAp is probably in the 5' portion of the mRNA, although it is definitely not part of the cap.

With the preliminary characterization of methyl group content and distribution accomplished, we have turned our attention to the examination of the methyl content of HnRNA. The preliminary results indicate that in HnRNA, >30S, there is about 6-8 times more N-6 methyl adenylic acid than in cap structures. In addition there is an appreciable amount of methyl-labeled T2 RNase resistant di- and trinucleotide material which is absent in mRNA. These methylated oligonucleotides could represent 2'-O-methylations at internal positions in the HnRNA which after cleavage present sites for the capping reaction as originally suggested by Rottman, Perry and Shatkin (1974). In this connection, it is pertinent to note that Abraham et al., (1975) have shown that the capping reaction with VSV, in contrast to the original reactions described for CPV and reoviruses, involves the addition of 5'ppm^{7}G from GTP to a 5' pA^{m}pXp – virion polymerase product (Fig. 8). Thus if an enzymatic cleavage in HnRNA left a 5' phosphate, a capping reaction with a diphosphate transfer from GTP might complete the reaction in a manner analogous to VSV capping (Fig. 9). Already the results with methylation of both HnRNA and mRNA are at least compatible with, if not strongly suggestive of, processing of HnRNA to mRNA being a major pathway for mRNA biosynthesis.

Obviously further work on the capping reaction and the sequences containing methyl groups in mRNA and HnRNA may add considerable additional

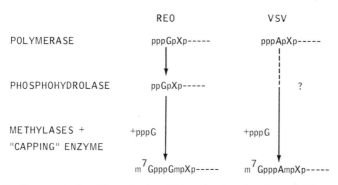

Fig. 8. *Two methods of "capping" mRNA molecules. Reovirus mRNA receives only a monophosphate from GTP during capping-methylation reaction. VSV mRNA receives a diphosphate during its capping reaction (See Furuichi et al., 1975a and Abraham et al., 1975). In the case of Reovirus the "capping" pppG adds a monophosphate, in the case of VSV the pppG adds a diphosphate.*

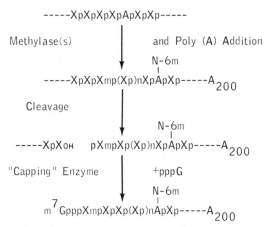

Fig. 9. *Proposed method of HnRNA cleavage and mRNA capping. Following methylation of the 2' OH groups of ribose moieties the HnRNA chain could be cleaved leaving a terminal t' phosphate. Completion of the 5' terminal "cap" structure could then be mediated by methylation. Capping enzyme(s) resulting in the transfer of a diphosphate from GTP.*

weight to the growing proof of sequence overlap between HnRNA and mRNA.

C. Evidence for the Precursor Nature of HnRNA.

As was mentioned early in this paper, proof that large HnRNA molecules are precursors of mRNA cannot rest only on the demonstration that sequences found in mRNA also are found in HnRNA but that HnRNA is the direct product of synthesis and mRNA is derived by post-transcriptional changes. Even though it has been recognized for years that most of the briefly labeled HnRNA products were larger than mRNA (Scherrer and Marcaud, 1965; Soeiro et al., 1966; Houssais and Attardi, 1966) — there was some labeled nuclear RNA of the size of mRNA and it remained possible that all the very large material turned over while the shorter molecules might be direct transcripts which were the only molecules destined to become mRNA. In addition a technical controversy regularly reappears over whether HnRNA really represents whole molecules or artificial aggregates of smaller molecules in spite of many demonstrations that heat, DNase, Me_2SO or other denaturing agents leave many long molecules intact as tested both by sedimentation analysis and EM visualization (Scherrer et al., 1963; Granboulan and Scherrer, 1969; Scherrer et al., 1970; Holmes and Bonner, 1972).

Recent experiments involving very brief times of nucleoside incorporation contribute further evidence that HnRNA is real and not aggregates of smaller molecules; further analysis of these brief label experiments also suggests an obligatory flow of labeled nucleosides through larger HnRNA before appearance of label in lower molecular weight mRNA.

Derman has examined the sedimentation profiles of native HnRNA labeled for 100 sec with [^{3}H]uridine and 100 minutes with [^{14}C]uridine. In addition the sedimentation behavior of molecules containing labeled poly(A) after a 200 minute exposure to ^{32}P and a 100 sec. exposure to [^{3}H]adenosine was also determined. In the HnRNA labeled with uridine (^{3}H and ^{14}C) the newest molecules sedimented slower than the steady state labeled molecules (Fig. 10). In contrast, the newest poly(A) containing molecules had a faster sedimentation profile than the poly(A) containing molecules present at steady state. These two results argue strongly against appreciable non-specific aggregation of HnRNA, since differentially marked molecules sediment differently. In addition, previous attempts to reconstruct aggregates by adding labeled smaller mRNA molecules containing or lacking poly(A) during the digestion and extraction of the nuclei have consistently failed to demonstrate

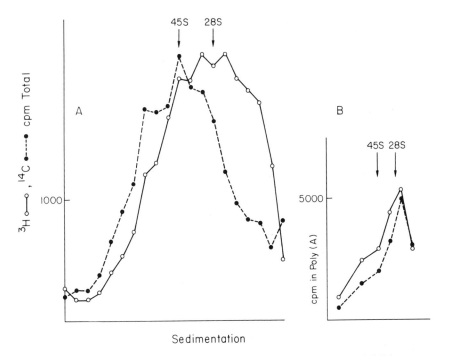

Fig. 10. *Sedimentation of nascent and steady state HnRNA. (A) HnRNA was prepared from cells (pretreated with 0.04 µg actinomycin) labeled for 100 sec with [^{3}H]uridine or for 100 min with [^{14}C]uridine and sedimented without denaturation through a sucrose-SDS gradient (See Fig. 2).*

(B) HnRNA fractionated similarly to (A) from cells labeled with [^{3}H] and [^{14}C]adenosine (100 sec and 100 min respectively) was analyzed in the six indicated regions for newly labeled and "steady-state" labeled poly(A) (−○− ^{3}H; −●− ^{14}C).

aggregation (Derman and Darnell, 1974). EM pictures showing long molecules and the presence of Me_2SO-stable, long polynucleotides have previously established the existence of long molecules; we believe this new evidence rules out the possibility of substantial aggregation under the conditions we use to study HnRNA in cultured cells.

In the experiment described above, the slower sedimentation of the [^{3}H]uridine (short label) compared to [^{14}C]uridine (long label) possibly signifies an equally labeled segment of RNA in growing chains of all sizes, demonstrating the size distribution of nascent RNA molecules, while the accumulated label at steady state indicates the minimal average size which the nascent RNA eventually achieves. If this is so then the great majority of HnRNA achieves a final chain size considerably greater than 5100 KB (the size of 28S rRNA) suggesting no significant amount of direct transcript in the size range of mRNA. In addition, other studies of Derman indicate that the short molecules to which poly(A) is attached in the nucleus become completely labeled in their internal chains only after cells are exposed to [^{3}H]adenosine for about 20 minutes. Thus, the increased level of radioactivity in short molecules in short labels demonstrated in Figure 10 is *not* into poly(A) containing molecules. The lengthy period necessary to label the non-poly(A) portion of the short poly(A) containing molecules is about equal to the time necessary for the very longest poly(A) terminated HnRNA (15-25 KB) to become completely labeled. These results suggest that the shorter RNA segments to which poly(A) is added, may be originally synthesized as part of much longer molecules and only after cleavage is poly(A) added. Experiments with total HeLa cell HnRNA and mRNA unfortunately can only suggest the precursor:product relationship, they do at the same time contribute negative evidence against the possibility of direct short transcripts being the most frequent mode of forming mRNA. The experiment to settle the precursor: product issue would necessarily include the demonstration that initially all of a specific mRNA region transcribed was part of a larger molecule(s) which in time yielded mRNA of the proper size. Two problems intervene in a simple solution of this problem using the popular recent technique of labeled cDNA made from a specific mRNA to probe for HnRNA that is complementary to the cDNA. (McKnight, G. S. and Schimke, 1974; Imaizumi et al., 1973). A cDNA probe will hybridize to all sequences to which it is complementary and if processed products have accumulated, then the use of radioactive cDNA as the probe might not detect the initial RNA product of a given gene. Therefore newly labeled RNA, labeled as close to the synthesis time of an RNA molecule as possible, must be examined for its content of specific sequences. This could be accomplished theoretically if large amounts of unlabeled cDNA were available. A second problem is that the amount of newly labeled RNA specific for one gene is very small (e.g., estimates for hemoglobin in erythroblasts is

10^{-4} of the total) and therefore enormous amounts of radioactive RNA precursors would be necessary to sufficiently label newly synthesized RNA in order to detect the gene specific RNA.

In order to circumvent some of these difficulties and at least determine whether in one case, high molecular weight RNA is the obligatory mode of synthesis for mRNA molecules, we have returned to a study of adenovirus mRNA formation. Adenovirus DNA is 39 KB and adeno-mRNA formation in infected cells seems to mimic cellular mRNA formation (Parsons et al., 1971). The DNA is transcribed in the nucleus, very high molecular weight adeno-specific "HnRNA" exists, (Parsons et al., 1971; Wall et al., 1972; Bachenheimer and Darnell, unpublished data) the mRNA is transported to the cytoplasm only if poly(A) attachment is allowed just as in the case of cellular mRNA (Philipson et al., 1971). In addition methyl groups are present in adeno-mRNA distributed between N-6 mAp and the blocked termini — "caps" — in the same ratio as in the cell mRNA (Sommer et al., unpublished data) suggesting each adeno-mRNA may also contain one cap structure and one N-6 mAp residue.

Adenovirus-specific RNA synthesis late in infection (after 15 hours) approaches 10% of the total nuclear RNA synthesis and is mainly from one strand (Philipson et al., pers. comm.). Because adenovirus-specific RNA represents such a large proportion of total RNA synthesis, label times as short as 1-2 minutes allow adenovirus-specific RNA to be measured. This exposure time is equal to the synthesis of polio RNA, 8 KB (Darnell et al., 1967) and about ½ of the 45S r-pre-RNA, 14KB (Greenberg and Penman, 1968). Furthermore, since adenovirus DNA can be specifically fragmented with restriction endonucleases (Peterson et al., 1973), the amount of briefly labeled RNA from each segment can be measured. Thus, an experiment modeled after the classic Dintzis experiment (1961) on chain growth can be performed with adenovirus RNA synthesis. If RNA chain growth begins at a single point and proceeds to the end of the DNA molecule, then very briefly labeled nascent RNA of increasing sedimentation size will be complementary to DNA fragments that lie at increasing distances from the point of origin (Fig. 11). By this test it has indeed been found that the only short labeled adenovirus-specific RNA after 1 min is complementary to the A fragment, which is at one end of the adenovirus genome. Proceeding to the other end of the genome fragment B is encountered before fragments E and C. The labeled RNA complementary to B is all larger than 15,000 nucleotides and the RNA complementary to E and C is still larger (sediments faster) than the B-specific RNA. After only two minutes of label, RNA that hybridized to B and E as well as A is found which is discrete in size and much smaller than the original very long molecules. This small RNA is approximately the same size as the cytoplasmic mRNA found after label times of several hours.

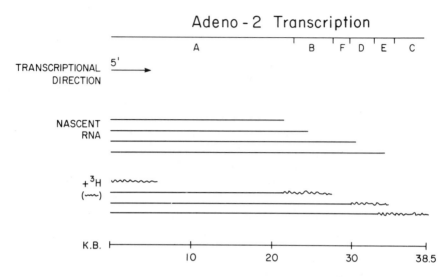

Fig. 11. *Model for Adeno-2 transcription with a single initiation site.*

The results strongly indicate at least for adenovirus mRNA production, the first gene products are very long molecules which can be processed very rapidly perhaps within a minute or two of synthesis. If this interpretation proves correct then it will be exceedingly difficult to investigate inside cells the true transcription product of cell genes.

III. CONCLUSION

A large body of evidence suggests sequences are shared between HnRNA and mRNA in both cultured cells and some tissues. These sequences include regions of RNA that encode protein sequence as well as poly(A) and methyl groups which are added after transcription. The difficulties of proving the true precursor nature of HnRNA have been described, as has an experiment which shows that adenovirus mRNA likely does go through an obligatory large molecule stage during its production.

Do we infer that all mRNA is formed via this HnRNA → mRNA pathway? Clearly we do not mean to imply this, although it does appear that a majority of mRNA in cultured cells might arrive via the HnRNA → mRNA pathway (Fig. 12). Other work indicates that for several specific mRNA molecules (interestingly all coding for proteins that are to be produced and excreted in massive amounts by their specific differentiated cells), the transcript size is very close to or the same as the final size (Fig. 12). These include the mRNA for silk fibroin of (Suzuki and Brown, 1972; Lizzardi et al., 1975; Lizzardi, pers. comm.) *Bombyx mori* and the major salivary gland product of

Possible modes of mRNA formation in mammalian cells

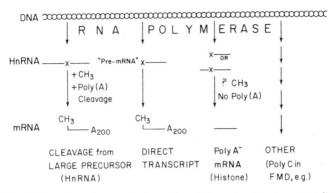

Fig. 12. *Possible modes of transcription and mRNA production in mammalian cells.*

Chironomus tentans (Daneholt, 1975). Both have been studied as labeled RNA molecules and both are in excess of 15,000 nucleotides in length. They apparently function as such in the manufacture of protein, so they are unusually large mRNA's to begin with and cleavage of a short region from a precursor could not easily be detected. It is almost certain, however, that they are never part of substantially longer molecules because nascent specific molecules which are shorter have been observed and never any specific RNA which is longer.

Recent claims have appeared that ovalbumin mRNA from chicken oviducts also follow this rule of transcriptional size equalling mRNA size (McKnight and Schimke, 1974). This conclusion is considerably less convincing than the two insect cases, because it was reached by hybridizing labeled cDNA to whole tissue RNA, where the completed mRNA itself must be the predominant class of molecules containing ovalbumin sequences. A careful titration was made of stable high molecular weight RNA and no very high molecular weight RNA complementary to ovalbumin cDNA was detected. However, if processing were very rapid and perhaps even coincident with transcription, then ovalbumin sequence could never be detected as part of larger molecules.

Pertinent to the question of the speed of processing are the adenovirus results described in this paper and the recent work of Laird and colleagues (pers. comm.). They have examined, by electron microscopy, what they believe to be growing chains of HnRNA in Drosophila embryonic cells. Two classes of molecules are seen a) those which reach perhaps 20-40 KB in length and b) those which derive from contiguous regions of DNA 20-40 KB long, but where the transcripts never exceed perhaps 5 KB. They believe this second class represents processing during transcription of some HnRNA molecules. Finally it must be stated, however, that direct transcription and utilization of some small mRNA's remains an open possibility.

In addition all of the above discussion has concerned mRNA molecules bearing poly(A) at their 3' end. What about molecules that lack poly(A)? Certain virus mRNAs and histone mRNA (Stoltzfus et al., 1973; Adesnik and Darnell, 1972; Greenberg and Perry, 1972) appear to lack poly(A) completely. Recently evidence for a class of cellular mRNA's lacking poly(A) has appeared (Milcarek et al., 1974). The mode of transcription of these mRNA's is totally unknown at present — they may or may not have higher molecular weight precursors.

In conclusion, many specific post-transcriptional steps occur in mRNA manufacture, probably including specific cleavage; in addition, different modes of transcription appear possible. As was stated at the outset, a final goal of this work is to locate the points at which regulation of mRNA supply occur and the knowledge of details of mRNA manufacture will certainly be helpful in eventually achieving that goal.

Abbreviations

1. HnRNA = heterogeneous nuclear RNA, the high molecular weight (5-50 KB) nuclear RNA fraction found in all higher cells.
2. KB = kilobases; for single strand nucleic acids = number of residues; for double-stranded nucleic acids = base pairs.
3. N p = 2'OH ribose methylated ribonucleotide.
4. N-6^{m}Ap = adenylic acid methylated in 6-NH_2 group of adenine.
5. m^7Gp = guanylic acid methylated in 7 position of purine ring.

ACKNOWLEDGEMENTS

1. This work was supported by grants from the National Institutes of Health (CA 16006-02), the National Science Foundation (GB 44016), and the American Cancer Society (VC 101D).
2. Larry Puckett is a Damon Runyon Fellow.
3. Pre-doctoral trainee (5-Tol-GM-02012-05).

REFERENCES

Abraham, G., Rhodes, D. P. and Bannerjee, A. K. (1975). *Cell,* in press.
Adesnik, M. and Darnell, J. E. (1972). *J. Mol. Biol.* **67**, 397–406.
Adesnik, M., Salditt, M., Thomas, W. and Darnell, J. E. (1972). *J. Mol. Biol.* **71**, 21–30.
Bernhardt, D. and Darnell, J. E. (1969). *J. Mol. Biol.* **42**, 43–56.
Britten, R. J. and Kohne, D. E. (1968). *Science* **161**, 529.
Burdon, R. H., Martin, B. T. and Lal, B. M. (1967). *J. Mol. Biol.* **28**, 357–371.
Daneholt, B. (1975). *Cell* **4**, 1–9.
Darnell, J. E., Girard, M., Baltimore, D., Summers, D. F. and Maizel, J. (1967). *In* "The Molecular Biology of Viruses" (J. Colter ed.), (N.Y. Acad. Press).

Darnell, J. E., Philipson, L., Wall, R. and Adesnik, M. (1971b). *Science* **174**, 507–510.
Darnell, J. E., Wall, R. and Tushinski, R. J. (1971a). *Proc. Nat. Acad. Sci. U.S.A.* **68**, 1321–1325.
Darnell, J. E., Jelinek, W. R. and Molloy, G. R. (1973). *Science* **181**, 1215–1221.
Derman, E. and Darnell, J. E. (1974). *Cell*, **3**, 255–264.
Dintzis, H. (1961). *Proc. Nat. Acad. Sci. U.S.A.* **47**, 247–261.
Diez, J. and Brawerman, G. (1974). *Proc. Nat. Acad. Sci. U.S.A.* **71**, 4091–4095.
Edmonds, M. P., Vaughan, M. H. and Nakazato, H. (1971). *Proc. Nat. Acad. Sci. U.S.A.* **68**, 1336–1340.
Furuichi, V. and Miura, K.-I. (1975a). *Nature* **253**, 347.
Furuichi, Y., Morgan, M., Muthurkrishivam, S. and Shatkin, A. J. (1975b). *Proc. Nat. Acad. Sci. U.S.A.* **72**, 362–366.
Furuichi, Y., Morgan, M., Shatkin, A. J., Jelinek, W., Salditt-Georgieff, M. and Darnell, J. E. (1975c). *Proc. Nat. Acad. Sci. U.S.A.* **72**, 1904–1908.
Georgieff, M., Bachenheimer, S. and Darnell, J. E. (1974). *Cold Spring Harbor Symp. Quant. Biol.* **39**, 475–482.
Granboulan, N. and Scherrer, K. (1959). *Eur. J. Biochem.* **9**, 1–20.
Greenberg, H. and Penman, S. (1966). *J. Mol. Biol.* **21**, 527–535.
Greenberg, J. R. and Perry, R. P. (1972). *J. Mol. Biol.* **72**, 91–98.
Greenblatt, J. and Schleif, R. (1971). *Nature New Biol.* **223**, 166.
Holmes, D. S. and Bonner, J. (1973). *Biochemistry* **12**, 2330–2338.
Houssais, J. F. and Attardi, G. (1966). *Proc. Nat. Acad. Sci. U.S.A.* **56**, 616–623.
Imaizumi, T., Diggelman, H. and Scherrer, K. (1973). *Proc. Nat. Acad. Sci. U.S.A.* **70**, 1122–1126.
Jacob, F. and Monod, J. (1961). *J. Mol. Biol.* **3**, 318–356.
Jelinek, W., Adesnik, M., Salditt, M., Sheiness, D., Wall, R., Molloy, G., Philipson, L. and Darnell, J. E. (1973a). *J. Mol. Biol.* **75**, 515–532.
Kacian, D. L., Spiegelman, S., Bank, A., Terada, M., Metafora, S., Dow, L. and Marks, P. A. (1972). *Nature* **235**, 167–169.
Kates, J. (1970). *Cold Spring Harbor Symp. Quant. Biol.* **35**, 743–752.
Lanyon, W. G., Ottolenghi, S. and Williamson, R. (1975). *Proc. Nat. Acad. Sci. U.S.A.* **72**, 258–262.
Lavi, S. and Shatkin, A. (1975). *Proc. Nat. Acad. Sci. U.S.A.* in press.
Lee, Y., Mendecki, J. and Brawerman, G. (1971). *Proc. Nat. Acad. Sci. U.S.A.* **68**, 1331–1335.
Lizzardi, P. M., Williamson, R. and Brown, D. D. (1975). *Cell* **4**, 199–205.
Lim, L. and Canellakis, E. S. (1970). *Nature* **227**, 710–712.
Levinthal, C., Fan, D., Higa, A. and Zimmerman, R. A. (1963). *Cold Spring Harbor Symp. Quant. Biol.* **28**, 183–190.
Lindberg, U. and Darnell, J. E. (1970). *Proc. Nat. Acad. Sci. U.S.A.* **65**, 1089–1096.
Maden, B. E. H. (1971). *Progr. Biophys. Mol. Biol.* **22**, 129.
McNaughton, M., Freeman, K. B. and Bishop, J. D. (1974). *Cell* **1**, 117–125.
McKnight, G. S. and Schimke, R. T. (1974). *Proc. Nat. Acad. Sci. U.S.A.* **71**, 4327–4331.
Melli, M. and Pemberton, R. (1972). *Nature New Biology* **236**, 172–174.
Mendecki, J., Lee, Y. and Brawerman, G. (1972). *Biochemistry* **11**, 792–798.
Milcarek, C., Price, R. P. and Penman, S. (1974). *Cell* **3**, 1–10.
Molloy, G. R., Sporn, M., Kelley, D. E. and Perry, R. P. (1972). *Biochemistry* **11**, 3256–3260.
Penman, S., Rosbash, M. and Penman, M. (1970). *Proc. Nat. Acad. Sci. U.S.A.* **67**, 1878–1885.

Perlman, S., Abelson, A. and Penman, S. (1973). *Proc. Nat. Acad. Sci. U.S.A.* **70**, 350–355.
Perry, R. P. (1962). *Proc. Nat. Acad. Sci. U.S.A.* **48**, 2179–2186.
Perry, R. P., Kelley, D. and LaToree, J. (1974). *J. Mol. Biol.* **82**, 315–331.
Perry, R. P. and Kelley, D. E. (1974). *Cell* **1**, 37–42.
Petersson, U., Mulder, C., Delius, H. and Sharp, P. A. (1973). *Proc. Nat. Acad. Sci. U.S.A.* **70**, 200–204.
Philipson, L., Wall, R., Glickman, R. and Darnell, J. E. (1971). *Proc. Nat. Acad. Sci. U.S.A.* **68**, 2806–2813.
Puckett, L., Chamber, S. and Darnell, J. E. (1975). *Proc. Nat. Acad. Sci. U.S.A.* **72**, 389–393.
Rottman, F., Shatkin, A. J. and Perry, R. P. (1974). *Cell* **3**, 197–199.
Robertson, H. and Dickson, L. (1974). *Brookhaven Symp. Biol.* **27** in press.
Ross, T., Aviv, H., Scolnick, E. and Leder, P. (1972). *Proc. Nat. Acad. Sci. U.S.A.* **69**, 264–268.
Roizman, B., Bachenheimer, S., Wagner, E. K. and Savage, T. (1970). *Cold Spring Harbor Symp. Quant. Biol.* **35**, 753–771.
Ruiz-Carillo, A., Beato, M., Schurtz, G., Feigelson, P. and Allfrey, V. G. (1973). *Proc. Nat. Acad. Sci. U.S.A.* **70**, 3641–3645.
Sambrook, J., Westphal, H., Srinivasan, P. R. and Dulbecco, R. (1968). *Proc. Nat. Acad. Sci. U.S.A.* **60**, 1288–1295.
Scherrer, K. and Darnell, J. E. (1962). *Biochem. Biophys. Res. Comm.* **7**, 486–490.
Scherrer, K., Latham, H. and Darnell, J. E. (1963). *Proc. Nat. Acad. Sci. U.S.A.* **49**, 240–248.
Scherrer, K. and Marcaud, L. (1965). *Bull. Soc. Chim. Biol.* **47**, 1697–1713.
Scherrer, K., Marcaud, L., Zajdela, F., London, I. M. and Gros, F. (1966). *Proc. Nat. Acad. Sci. U.S.A.* **56**, 1571–1578.
Scherrer, K., Spohr, G., Granbaulan, N., Morel, C., Grosclaude, J. and Chezzi, C. (1970). *Cold Spring Harbor Symp. Quant. Biol.* **35**, 539–554.
Shearer, R. N. and McCarthy, R. J. (1967). *Biochemistry* **6**, 283–289.
Shearer, R. W. and McCarthy, B. J. (1970). *J. Cell Phys.* **75**, 97–106.
Sheldon, R., Kates, J., Kelley, P. and Perry, R. P. (1972). *Biochemistry* **11**, 3829–3834.
Soeiro, R., Birnboim, H. C. and Darnell, J. E. (1966). *J. Mol. Biol.* **19**, 362–372.
Soeiro, R. and Darnell, J. E. (1970). *J. Cell Biol.* **44**, 467–475.
Sommers, S., Bachenheimer, S., Salditt-Georgieff, M., Jelinek, W. Darnell, J. E., Furuichi, H. and Shatkin, A. T. (1975). In preparation.
Stoltzfus, C. M., Shatkin, A. J. and Bannerjee, H. K. (1973). *J. Biol. Chem.* **248**, 7993–7998.
Suzuki, Y. and Brown, D. D. (1972). *J. Mol. Biol.* **63**, 409–429.
Tonegawa, S., Walter, G., Bernardini, A. and Dulbecco, R. (1970). *Cold Spring Harbor Symp. Quant. Biol.* **35**, 823–831.
Verma, I. M., Temple, G. F., Fan, H. and Baltimore, D., (1972). *Nature* **235**, 163.
Wagner, E., Penman, S. and Ingram, V. (1967). *J. Mol. Biol.* **29**, 371–387.
Wall, R., Weber, J., Gage, Z. and Darnell, J. E. (1973). *J. Virol.* **11**, 953–960.
Wall, R., Philipson, L. and Darnell, J. E. (1972). *Virology* **50**, 27–34.
Wei, C. W. and Moss, B. (1975). *Proc. Nat. Acad. Sci. U.S.A.* **72**, 318–322.
Whiteley, A. H., McCarthy, B. J. and Whiteley, H. R. (1966). *Proc. Nat. Acad. Sci. U.S.A.* **55**, 519–525.
Zubay, G., Schwartz, D. and Beckwith, J. (1970). *Proc. Nat. Acad. Sci. U.S.A.* **66**, 104–110.

SYNTHESIS AND TRANSLATION OF MESSENGER RNA DURING DIFFERENTIATION OF THE CELLULAR SLIME MOLD *DICTYOSTELIUM DISCOIDEUM*

Harvey F. Lodish, Tom Alton, Robert P. Dottin,
Alan M. Weiner and Jeanne P. Margolskee

*Department of Biology,
Massachusetts Institute of Technology,
Cambridge, Massachusetts 02139*

I. INTRODUCTION

Progress in understanding the complex molecular events which regulate differentiation in eukaryotic cells will require the sustained and collaborative efforts of many research groups working on a few advantageous systems. It seems clear that a wide variety of genetic, biochemical, and morphologic approaches will have to be used and that no one organism or system will be ideal for all these types of studies.

The cellular slime mold *Dictyostelium discoideum* offers many advantages for a detailed study of differentiation in a relatively simple system. Recent reviews on *Dictyostelium* have been written by Ashworth (1971), Bonner (1971), Jacobson and Lodish (1975), Killick and Wright (1974), Loomis (1975), Newell (1971), and Robertson and Cohen (1972). *Dictyostelium* is a eukaryotic protist easily cultivated as single cells either axenically or in association with bacteria (Sussman and Sussman, 1967; Watts and Ashworth, 1970). The nuclear DNA genome is one fiftieth the complexity of that of mammalian cells (Firtel and Bonner, 1972), and, as we shall discuss, the mode of messenger RNA biosynthesis is conservative and involves a much shorter nuclear precursor than in mammalian cells (Firtel and Lodish, 1973; Lodish, et al., 1973, 1974); these attributes facilitate labeling and RNA:DNA hybridization studies (Firtel and Lodish, 1973; Firtel, Jacobson, and Lodish, 1972). Isolated cells will form clones with an efficiency of nearly 100%. Common strains of *Dictyostelium* are haploid, and a large variety of mutations that affect growth or differentiation can be easily isolated (Loomis, 1969). Genetic analysis is possible although difficult. Parasexual genetics involves the rare fusion of two haploid strains to yield a diploid, followed by the rare production of haploid segregants. It has already been used to map 19 mutations into 5 linkage groups (Kessin, et al., 1974; Katz and Kao, 1974; Williams, et al., 1974; Gingold and Ashworth, 1974). The recent discovery of a presumptive sexual cycle and meiosis could enormously facilitate genetic studies (Erdos, et al., 1973; Clark, et al., 1973; Erdos, et al., 1975).

It is the well defined morphogenetic sequence outlined in Figure 1, however, which has made *Dictyostelium* a prime system for a detailed study

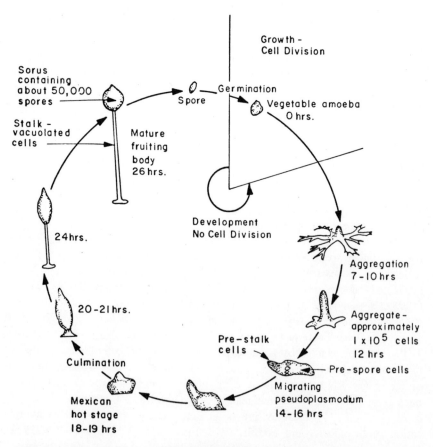

Fig. 1. *Life cycle of Dictyostelium discoideum. This figure depicts the commonly studied aspects of the slime mold life cycle: vegetative growth of haploid amoebae and formation of fruiting bodies. Not shown here is the formation and germination of macrocysts which is reviewed by Jacobson and Lodish (1975).*

of the molecular basis of differentiation. The entire developmental sequence can be studied under rigorously defined experimental conditions with sufficient numbers of cells to permit most conventional biochemical procedures. Morphogenesis occurs only after growth has ceased and occurs in the absence of DNA replication and other biochemical activities characteristic of growing cells (Katz and Bourguignon, 1974).

The developmental phase is triggered by starvation (Sussman, 1966). Amoebae growing in association with bacteria on an agar plate will initiate development when the bacterial food is exhausted. Development can be made more synchronous if growing cells (either axenically or in association with bacteria) are removed from nutrients by centrifugation and resuspended in a solution containing only inorganic salts. They are then spread on a Millipore filter or on a disc of filter paper resting on thick absorbant pads. Alternatively, non-nutrient agar can be used.

After an interphase of six to seven hours, the cells gradually aggregate into mounds containing about 10^5 cells. This aggregation is a chemotactic response to $3':5'$-cyclic AMP (cAMP) (Bonner, 1971); it is dependent on the appearance of both excreted and surface components which are involved in the binding and degradation of cAMP and on the appearance of new surface proteins required for cell adhesion (reviewed in Jacobson and Lodish, 1975; Gerisch, et al., 1974).

These mounds gradually form the pseudoplasmodium, a stage capable of migration. This "slug" stage is both thermotactic and phototactic (Bonner, et al., 1950). At this stage, one can observe differences in what was initially a homogeneous cell population. Microdissection studies showed that the rear two-thirds of the cell mass is committed (but reversibly) to differentiate into spore cells (Raper, 1940); these prespore cells contain a characteristic prespore vesicle (Hohl and Hamamoto, 1969). The front third of the mass is committed to differentiate into stalk cells. The tip of the pseudoplasmodium, comprising the front 10% of the cells, acts as an embryological organizer and is essential for proper differentiation; it may act by secreting cAMP continuously (Durston, 1974). None of the cells in the aggregate fuse, and cells can be mechanically separated at any stage.

The culmination phase of differentiation is induced either by a high ionic strength medium or by overhead light. The pseudoplasmodium rises apically and forms a flattened disc resembling a Mexican hat. The prestalk cells are found in the center of the aggregate. They develop into vacuolated stalk cells that extend lengthwise, causing a rapid elongation of the stalk. The spore mass is pushed upwards by the elongating stalk and mature into a spore mass surmounting a cellulose-encased rod of dead cells.

At specific stages during morphogenesis, the specific activities of many enzymes increase considerably. In several cases there is a subsequent decrease in enzyme activity. For at least one of these enzymes the increase in specific activity can be accounted for by *de novo* synthesis of enzyme protein (Franke and Sussman, 1973). Recent studies in which a combination of RNA synthesis inhibitors are used suggest that, in contrast to earlier studies, there is only a short lag between synthesis and translation of mRNAs encoding these developmentally regulated enzymes (Firtel, Baxter, and Lodish, 1973). Further, as we discuss below, experiments in which cells have been pulse-labeled with radioactive amino acids at discrete stages of the life cycle have shown that a number of polypeptides (identified as "bands" on polyacrylamide gels) are preferentially labeled at specific developmental stages (Tuchman, *et al.*, 1974). The specific labeling patterns are qualitatively similar even in very short pulses of amino acids (Alton and Lodish, manuscript in preparation). These results strongly suggest that selective protein turnover is not the mechanism responsible for the observed changes.

Additional results which demonstrate changes in gene transcription come from RNA:DNA hyribdization experiments. These showed that approximately half of the non-reiterated portion of the genome is transcribed throughout the entire course of the life cycle. Of the RNA transcribed, half appears to be present in cells of all developmental stages, whereas half can be subdivided into stage-specific subclasses (Firtel, 1972).

Our own research is directed primarily toward the factors which regulate synthesis and translation of mRNA during differentiation. In this paper we summarize our recent work in these areas.

II. *DICTYOSTELIUM* MESSENGER RNA

Dictyostelium messenger RNA has many features in common with mRNA from lower eukaryotes such as HeLa cells (Darnell, *et al.*, 1973). The average *Dictyostelium* mRNA contains about 1300 nucleotides, large enough to encode a single protein of molecular weight 40,000. The bulk of the mRNA is transcribed from non-reiterated sequences in the DNA. Less than 10% of mRNA is complementary to (i.e., derived from) the 30% of nuclear DNA sequences — repetitive DNA — which are present in 100-300 copies per genome (Firtel, *et al.*, 1972). At the 3' end of most mRNAs is a sequence of polyadenylic acid about 100 bases long. As in mammalian cells, this sequence is added to the mRNA after it is transcribed from DNA. Each *Dictyostelium* mRNA contains also a second sequence of poly(A), this only 25 bases long. It is located near the 3' end but is separated from the poly(A)$_{100}$ by from 1 to 20 bases (Lodish, *et al.*, 1973; Jacobson, *et al.*, 1974).

The poly(A)$_{25}$, in contrast to the poly(A)$_{100}$ sequence, is believed to be transcribed from poly(T)$_{25}$ sequences in the *Dictyostelium* genome (Fig. 2); there are about 15,000 poly(dT)$_{25}$ sequences in *Dictyostelium* DNA — about one per gene (Lodish, *et al.*, 1973; Jacobson, *et al.*, 1974).

The experiment in Figure 3 shows that the poly(A)$_{100}$ sequences are added just before the RNA exits from the nucleus and also supports the notion that the poly(A)$_{25}$ sequences are transcribed from DNA: following

STRUCTURE AND TRANSCRIPTION OF DICTYOSTELIUM DNA

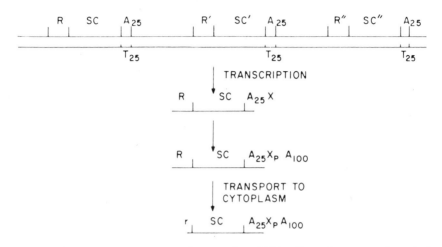

Fig. 2. *A model of Dictyostelium nuclear DNA. The primary genetic unit in Dictyostelium contains a repetitive DNA sequence (R) that averages 300 to 350 nucleotides at the 5' end; a sequence of nonreiterated or single-copy (SC) DNA of 1100 to 1200 nucleotides; and a sequence of 25 adenylic acid residues (A_{25}) at the 3' end. This unit is transcribed by RNA polymerase and then poly(A) of 100 to 150 residues is added after transcription. Before transport of the heterogeneous RNA to the cytoplasm, the majority of the 5' repetitive sequences are removed, leaving a short repetitive sequence. The evidence for the presence of a portion of the repetitive sequence transcript (r) on the majority of mRNA molecules is given in Lodish, et al. (1973). Since the poly(A)$_{25}$ and poly(A)$_{100}$ are separable on polyacrylamide gels after digestion with RNase T1 and RNase A, at least one other nucleotide (X) must be present between the 3' end of the short poly(A)$_{25}$ and the 5' end of the larger poly(A)$_{100}$ (Jacobson, et al., 1974).*

addition of [^{3}H]adenosine to vegetative cells, radioactive poly(A)$_{100}$ sequences, attached to mRNA, appear in the cytoplasm after only a 2 minute lag period. By contrast, labeled poly(A)$_{25}$ sequences and labeled nonpoly(A) segments of mRNA (i.e., cytoplasmic polyadenylated RNA) are found in the cytoplasm only after a 4 minute lag. These results mean that it takes only 4 minutes for an mRNA precursor molecule, containing a poly(A)$_{25}$ sequence, to be processed and be transported from the nucleus. The poly(A)$_{100-150}$ sequence is added during the final 2 minute period in the nucleus.

It is of interest to point out that mammalian mRNA does not contain poly(A)$_{25}$ sequences. Such sequences are, however, found in the large nuclear heterogeneous RNA, the presumed precursor to mammalian mRNA (Nakazato, et al., 1974).

Many, if not all, cellular and viral mRNAs in mammalian cells contain an unusual sequence of nucleotides at the 5' end (Adams and Cory, 1975;

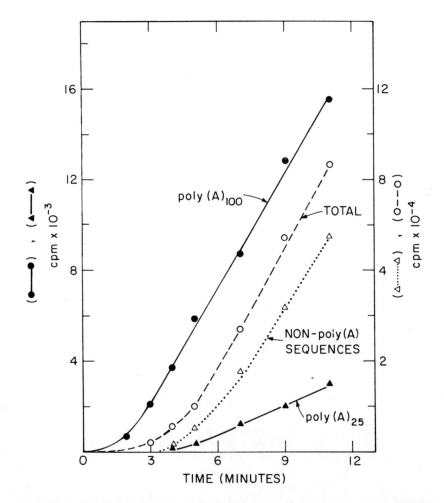

Fig. 3. *Kinetics of appearance of poly(A)$_{100}$ and poly(A)$_{25}$ in the cytoplasm. Vegetative cells were labeled with [^{3}H]adenosine; at various times cytoplasmic RNA was isolated by the proteinase K method (Firtel and Lodish, 1973). Poly(A)-containing RNA was purified by chromatography on poly(U)-Sepharose. Total poly(A) was determined by the fraction of the [^{3}H]adenosine label which was resistant to RNases A and T1; the fraction of poly(A) radioactivity which represents poly(A)$_{100}$ and poly(A)$_{25}$ sequences was determined by electrophoresis on 10% polyacrylamide gels as in Jacobson, et al. (1974b). Non-poly(A) sequences were the fraction of [^{3}H]-radioactivity which was sensitive to the RNases. All samples were normalized for a given amount of rRNA present in the extracted samples.* ○, *total poly(A)-containing RNA;* △, *non-poly(A) sequences;* ●, *poly(A)$_{100}$;* ▲ *poly(A)$_{25}$.*

Furuichi, et al., 1975a, 1975b; Perry, et al., 1975; Wei, et al., 1975). A general formula for these sequences is $m^7G^{5'}ppp^{5'}XmpYpZp$... A 7-methyl guanosine is linked by a $5':5'$-triphosphate to the nucleoside X at the $5'$ end of the RNA. X also contains a $2'0$-methyl substituent, and in many RNAs the second base, Y, also contains a $2'0$-methyl group. In mRNA from HeLa, L, or murine plasmacytoma cells, X can be any of the four nucleotides. In the case of reovirus or vesicular stomatitis virus, the 7-methyl guanosine and $2'0$-methyl groups are added by virion transcriptases, and the $5'$ nucleoside X is either A or G (Furuichi, et al., 1975a). Mammalian cell mRNA also contains one or more 6-methyl adenine residues, although the position of the mA residue in the mRNA is not clear (Furuichi, et al., 1975b; Perry, et al., 1975; Wei, et al., 1975).

It is of interest that *Dictyostelium* mRNA contains some, but not all, of the methyl groups characteristic of mammalian mRNA. Some of these results are shown by the experiments in Figure 4. $[^{32}P]$-labeled mRNA (cytoplasmic RNA purified by chromatography on oligo(dT)-cellulose) was digested with alkali and then chromatographed on a column of DEAE cellulose using a linear gradient of NaCl (0.05 M to 0.4 M) in 7 M urea. Over 99% of the label eluted, as expected, with mononucleotides containing a charge of -2. But a small fraction (0.5%) eluted at a position indicating it had 4 negative charges (Fig. 4A). A smaller fraction appears to have 5 negative charges. mRNA can also be labeled with [methyl-^{3}H] methionine (Fig. 4C); when [methyl-^{3}H]-labeled mRNA was digested with alkali and chromatographed, 60% of the material eluted at the -4 and -5 peaks coincident with the $[^{32}P]$ (compare Fig. 4A and 4C). Methyl-labeled rRNA yielded no such peak (Fig. 4B); all of the material eluted at -2 (mononucleotides) or -3 (dinucleotides containing, presumably, a $2'0$-methylated nucleotide). These and other results establish the absence of 6-methyl adenosine in *Dictyostelium* mRNA.

The oligonucleotides in the -4 and -5 peaks from mRNA can be resolved into three well-separated spots by electrophoresis on DEAE paper at pH 3.5 (Fig. 5). Up to 3% of the label remains at the origin. Recently we showed that the sequence of the oligonucleotide in the major spot (spot I, 65% of the total) is $m^7G^{5'}ppp^{5'}Ap$ and that of another spot (II, 10% of total) is most likely $m^7G^{5'}ppp^{5'}Gp$. These two oligonucleotides elute, as expected, at the -4 position of the DEAE column. The remaining 25% of the material in spot III (the -5 peak) appears to have the structure $m^7GpppXmpYp$; we do not yet know the identity of X or Y. Nonetheless, the strong bias towards A as the $5'$ nucleotide and the absence of $2'0$-methyl substituents are clear differences from mammalian mRNA. The recovery of spots I-III is such that each mRNA appears to contain one of the blocked nucleotides at the $5'$ end (Dottin, Weiner, and Lodish, in preparation).

The mechanism of biogenesis of messenger RNA in *Dictyostelium* appears to be much simpler than in mammalian cells. Nuclei of *Dictyostelium* cells

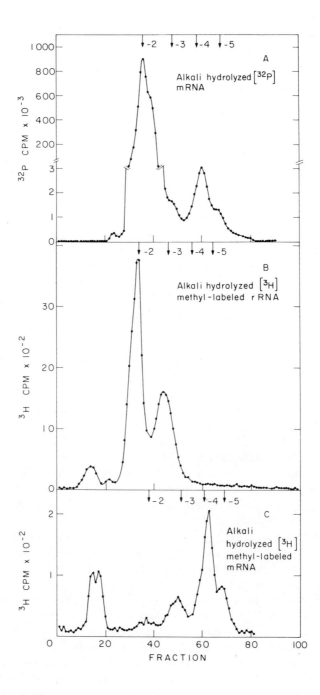

labeled *in vivo* or *in vitro* under a variety of conditions do not contain material analogous to the large nuclear heterogeneous RNA found in mammalian cells (Firtel and Lodish, 1973). Mammalian HnRNA is from 2-30 times the size of mRNA, and over 90% of the HnRNA is degraded within the nucleus. Apparently a small amount of HnRNA is the precursor to mRNA (Darnell, *et al.*, 1973). By contrast, the majority of pulse-labeled *Dictyostelium* nuclear RNA that is not a precursor to rRNA has an average weight of 500,000 daltons, only 20% longer than cytoplasmic mRNA (Firtel and Lodish, 1973). These RNAs contain at least the shorter sequences of polyadenylic acid, poly(A)$_{25}$ (Jacobson, *et al.*, 1974a, 1974b). Pulse-labeling experiments showed that at least 70% of the nuclear poly(A)-containing RNA is a material precursor to cytoplasmic mRNA (Firtel and Lodish, 1973; Lodish, *et al.*, 1973). This point was substantiated by RNA:DNA hybridization experiments using DNA complementary to mRNA synthesized by reverse transcriptase (Lodish, *et al.*, 1973; Verma, *et al.*, 1974). When intact cells are labeled for short periods of time, these nuclear poly(A)-containing RNA molecules do not contain a triphosphate group at the 5' terminus. However, full length poly(A)$_{25}$-containing pre-mRNAs can be synthesized in isolated nuclei, as shown in Figure 6 (Jacobson, *et al.*, 1974b), and an appreciable fraction of the labeled poly(A)-containing RNA molecules do possess a 5' triphosphate residue (A. Jacobson, personal communication). This is stong evidence that these nuclear poly(A)-containing RNAs represent a primary transcript of the nuclear DNA and are not derived from a longer RNA precursor.

Approximately 1200 of the 1500 bases in the average nuclear precursor to mRNA is complementary to, and hence transcribed from, single-copy DNA. Each molecule of pre-mRNA also contains at or near the 5' end about 300 nucleotides which are transcribed from repetitive DNA and contains near the 3' end a sequence of poly(A)$_{25}$ which is presumably transcribed from the poly(T)$_{25}$ sequences in the nuclear DNA (Firtel and Lodish, 1973; Jacobson,

Fig. 4. *(at left) Column chromatographic pattern of nucleotides from Dictyostelium RNA. Vegetative cells were starved for 2 hrs in suspension [2 X 10^8 cells in 20 ml MES-PDF (Firtel and Lodish, 1973)] prior to pulse labeling for 1 hr (30 mCi [^{32}P] phosphoric acid or 1 mCi [methyl-^{3}H] methionine). RNA from the cytoplasm was extracted with cold phenol, precipitated with ethanol, and applied to oligo(dT)-cellulose to purify poly(A)-containing mRNA from rRNA and tRNA. RNA was reprecipitated and then digested in alkali (20 μl 0.5 M NaOH for 16 hrs at 37°), neutralized, and chromatographed on 1 x 20 cm DEAE cellulose column with a gradient of NaCl (0.05 M to 0.4 M in 7 M urea, 50 mM Tris, pH 8.0). Yeast RNA digested with RNase A was included as marker in order to determine the positions of oligonucleotides with charges of −2 to −5. Fractions (2 ml) were collected and counted. (A) dT-binding [^{32}P] mRNA; (B) dT column flowthrough of [methyl-^{3}H] methionine-labeled RNA containing mainly rRNA and tRNA; (C) dT-binding [methyl-^{3}H] methionine-labeled RNA (mRNA).*

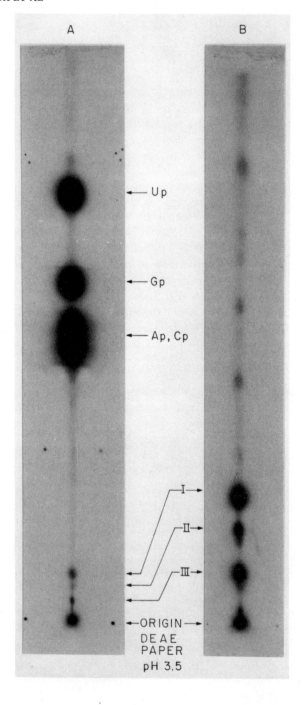

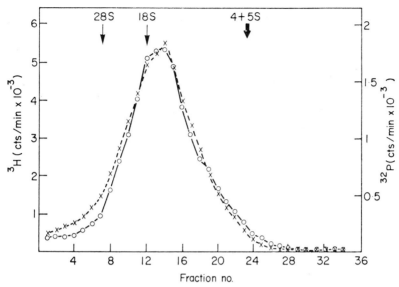

Fig. 6. *Dimethylsulfoxide-sucrose gradient analysis of mRNA precursor synthesized in isolated nuclei.* Messenger RNA precursor was labeled in isolated nuclei (Jacobson, et al., 1974a) or in whole cells and purified by poly(U)-Sepharose chromatography (Firtel, et al., 1972; Firtel and Lodish, 1973). In vitro labeling was with $[\alpha^{32}P]$ UTP; in vivo labeling was with $[^3H]$ uracil (Jacobson, et al., 1973; Firtel and Lodish, 1973). Aliquots of the $[^{32}P]$- and $[^3H]$-labeled molecules were mixed and centrifuged on a zero to 8% sucrose gradient containing 99% dimethylsulfoxide, 0.001 M sodium EDTA, and 0.00025 M TES, pH 6.8. Constant volume fractions were collected from the bottom of the tube after centrifugation. Radioactivity was determined by adding 1.5 ml water and 10 ml aquasol to each sample. (–X–X–), mRNA precursor labeled with $[^3H]$ uracil in whole cells; (–o–o–), mRNA precursor labeled with $[\alpha^{32}P]$ UTP in isolated nuclei.

et al., 1974a). The presence in the same primary RNA transcript of RNA sequences transcribed from repetitive DNA, single-copy DNA, and poly(T) is in agreement with the length and interspersion of these sequences within the nuclear DNA genome: approximately 60-70% of the *Dictyostelium* genome consists of interspersed reiterated and single-copy DNA sequences. The interpsered repetitive DNA sequences have an average length of 250-400 nucleotide pairs, whereas the average length of the interspersed single-copy

Fig. 5. *(at left) Separation of blocked 5' nucleotides by paper electrophoresis.* A sample of $[^{32}P]$-labeled mRNA (Fig. 4) was digested with a mixture of RNases A, T1, and T2. The resulting digest was analyzed by electrophoresis on DEAE paper at pH 3.5 (left panel). The electrophoresis in the right panel used more $[^{32}P]$ RNA and was done for a longer period of time, so that the mononucleotides ran off the end of the paper. The paper was exposed to Kodak Royal Blue X-ray film, and shown here is a photograph of this radioautogram.

sequences is between 1000 and 1200 nucleotide pairs (Fig. 2). [Approximately 20% of the single-copy DNA sequences is found in tracts 3000 nucleotides long without interspersed reiterated DNA sequences, and about 35% of the repetitive DNA sequences are also found in long tracts without interspersed single-copy DNA sequences (Firtel and Kindle, 1975).]

Nuclear pre-mRNAs appear to be subjected, within the nucleus, to two types of processing reactions before they exit into the cytoplasm as messenger RNA. It is not known in what order these occur, nor is it clear whether either is coupled to transport from the nucleus. First, about 300 bases — most of the transcripts from the repetitive DNA — appear to be removed from the 5' end. Second, a sequence of about 100-150 adenylic acid residues is added to the 3' end of the RNA (Lodish, *et al.*, 1973; Firtel and Lodish, 1973; Jacobson, *et al.*, 1974b).

In recent experiments we have discovered in nuclear polyadenylated RNAs the presence of 5' sequences containing an $m^7GpppXp$ residue; we are trying to understand the relationship of addition of the m^7G residue to mRNA processing. It is important to emphasize that our work, to date, deals only with mixed populations of mRNA and pre-mRNA. Our work by no means eliminates the possibility that some mRNA species are synthesized without any longer precursor at all.

III. PROTEIN SYNTHESIS DURING DIFFERENTIATION

The synthesis of none of the developmentally regulated enzymes studied to date represents more than 1% of the cell's total protein synthesis at any instant. It is also well known that enzyme activity is not necessarily a valid measure of the amount of enzyme protein. In any study of synthesis and translation of mRNA, it is easier to focus on those proteins (and presumably mRNAs) which are made in larger amounts; for this reason we decided to characterize the predominant species of polypeptides which are synthesized during different developmental stages (Tuchman, *et al.*, 1974). Figure 7A shows a radioautogram of an SDS polyacrylamide gel analysis of protein synthesized during 30 minute pulses of $[^{35}S]$methionine given at hourly intervals throughout development of wild-type cells.

As determined by the differences in the patterns of $[^{35}S]$methione incorporation into proteins at different times during development, the *Dictyostelium* morphogenetic sequence can be divided into at least four general stages: preaggregation (0 to 7 hours), aggregation (8 to 12 hours), postaggregation (13 to 18 hours), and culmination (19 to 24 hours). During the preaggregation stage, as much as 30% of the $[^{35}S]$methionine incorporated by the cells represents a single species which we have shown to be *Dictyostelium* actin.

As one example, Figure 8 shows that the predominant polypeptide synthesized during early development comigrates on SDS-polyacrylamide gels with authentic *Dictyostelium* actin. This labeled polypeptide also co-polymerizes with authentic actin and also contains the unusual amino acid 3-methylhistidine, characteristic of *Dictyostelium* and vertebrate actins.

The percentage of total cell protein synthesis which is actin rises from 8% in growing cells to over 30% at 3 hr of development (Fig. 9). During the aggregation stage, the synthesis of actin declines relative to total cell protein synthesis; only 7% of protein synthesized by 14 hr cells is actin, and this percentage drops further at later stages (c.f., Fig. 7A). During the aggregation stage, other major changes occur in the pattern of proteins synthesized. The synthesis of two other predominant "pre-aggregation" proteins decreases significantly, while the synthesis of at least four new "aggregation stage" polypeptides increases.

Much of our work to date has focused on the regulation of actin synthesis during differentiation. We have shown that *Dictyostelium* actin mRNA can be translated faithfully in cell-free extracts from rabbit reticulocytes or wheat germ, and we take the amount of actin synthesized in these extracts as a measure of the amount of actin mRNA (Alton and Lodish, in preparation). As an example, Figure 10 shows that poly(A)-containing RNA isolated from the cytoplasm of developing cells of age 2 hr directs the synthesis, in wheat germ extracts, predominantly of a single polypeptide which comigrates on SDS-polyacrylamide gel electrophoresis with *Dictyostelium* actin. Analysis of tryptic peptides of this *in vitro*-produced protein by paper electrophoresis shows that it is indeed actin (Fig. 11). As shown by the dashed lines in Figure 10, it appears that other authentic *Dictyostelium* polypeptides are made in this cell-free system, but we have not yet confirmed their identity by peptide mapping.

It is of interest that about one-quarter of the messenger activity in *Dictyostelium* cytoplasmic RNA does not stick to oligo(dT)-cellulose, even after several passages, and presumably does not contain a poly(A) sequence or contains one less than 20 bases long. This RNA does direct the cell-free synthesis of *Dictyostelium* actin and of the other major polypeptides whose synthesis is directed by poly(A)-containing RNA (Lodish, *et al.*, 1974). We suspect that these mRNAs are derived from poly(A)-containing RNAs whose sequences become short with age. We have no evidence for a discrete class of mRNAs which lack a poly(A) sequence such as has been reported in other eukaryotic systems (Milcarek, *et al.*, 1974). In experiments in which we wish to quantitate the total amount of actin (or other) mRNA activity, we utilize unfractionated cytoplasmic or whole cell RNA.

The amount of translatable actin mRNA at the different developmental stages, relative to the amount of rRNA, is shown in Figure 9 and Figure 12.

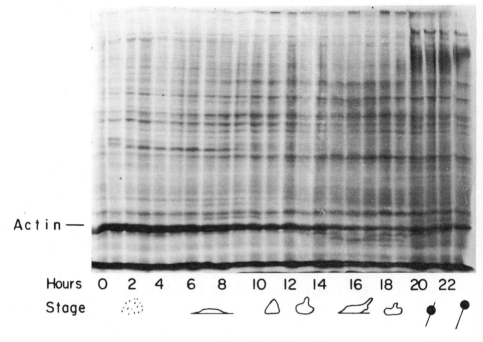
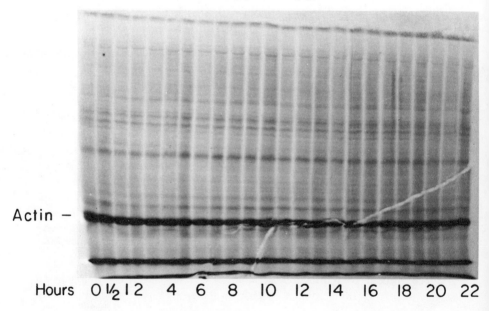

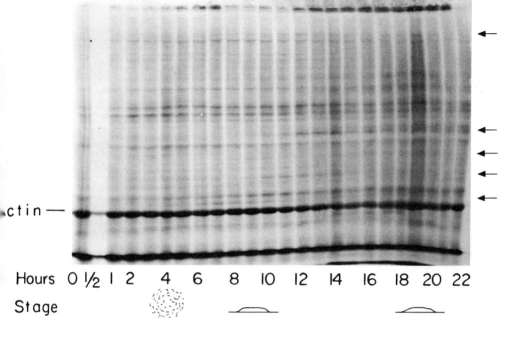

Fig. 7. *(left and top) Protein synthesis throughout development, in Dictyostelium. (A) wild-type AX 3 cells. (B) agg 2, a mutant unable to aggregate. (C) JM 94, a mutant able to aggregate but not form slugs.*

Exponentially growing cells were harvested and washed; 5×10^6 cells were plated on each of many 13 mm diameter black Millipore filters supported by Millipore pads saturated with PDF (Tuchman, et al., 1974). For labeling one filter was transferred to a 10 μl drop of [^{35}S]methionine solution (1 mCi per ml) in PDF and incubated 30 minutes. The incorporation was stopped by immersing the filter in Laemmli's sample buffer (Laemmli, 1970), vortexing, and boiling for 3 minutes. Aliquots were then electrophoresed through a 7.5% SDS polyacrylamide gel by the method of Laemmli (1970). An equal amount of acid insoluble radioactivity was applied to each slot in the slab. Cell streaming was visible at 8 hours, slug formation at 15 hours, and culmination was complete by 24 hours, in the case of AX3.

The mutant agg 2 (obtained from D. McMahon) was analyzed by the same method as was the wild-type with the addition of a labeling point between 0.5 to 1.0 hours.

The mutant JM 94 was induced in the strain AX3 by mutagenesis with N-methyl-N'-nitro-N-nitrosoguanidine and was analyzed by the same method as was the wildtype. This mutant shows cell streaming at 8 hours and forms discrete flat aggregates by 10 hours. These flat aggregates never differentiate further.

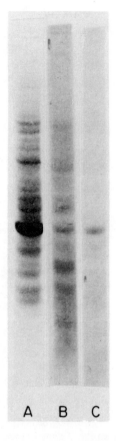

Fig. 8. SDS-polyacrylamide gels of various Dictyostelium samples. (A) Autoradiograph of 0 to 2 hr labeled sample. (B) Coomassie Blue stain of Dictyostelium membrane fraction from developing cells. (C) Coomassie Blue stain of purified Dictyostelium actin. Conditions for gel electrophoresis and for purification of actin were given in Tuchman, et al. (1975).

The amount of rRNA per cell gradually declines during differentiation; at 24 hr each cell contains only about one third the initial amount of RNA. Hence, the values in Figure 9 would have to be corrected in order to give the value of actin mRNA per cell. Nonetheless, it is clear that the amount of translatable actin mRNA per cell increases three-fold during the first three hours of development and then declines. The apparent half-life of actin mRNA after 3 hr is about 4-6 hr. It should be pointed out that we do not yet know the absolute rate of actin synthesis per cell, only the rate relative to other cell proteins. But Figure 9 does indicate that there is a good correlation between the relative rate of actin synthesis and the amount of translatable actin mRNA per cell.

Although direct evidence is lacking, we surmise that the increased level of translatable actin mRNA during the first three hours is a result of *de novo*

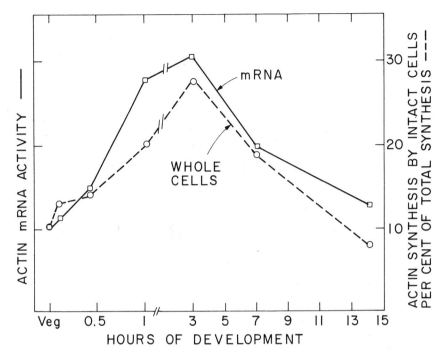

Fig. 9. *Content of actin mRNA and rate of actin synthesis. Actin synthesis by whole cells, (–○–). Radioautograms similar to those of Fig. 7A were scanned, and the fraction of the total area which was under the actin peak was determined. Content of actin mRNA, (–□–). The amount of actin mRNA, relative to ribosomal RNA, was estimated from the slopes of the curves in Fig. 12.*

synthesis of new actin mRNA. In growing cells very little of the polyadenylated RNA is found in the cytoplasm unattached to polysomes (see below), and little, if any, accumulates in the nucleus.

Further insight into the regulation of protein synthesis during the early phases of differentiation has come from a study of mutants which are unable to aggregate normally. Thus far, we have studied ten independently isolated aggregation-deficient mutants. These mutants either fail to aggregate at all or form flat, diffuse aggregates which never become spherical or progress into slugs.

Mutants which fail to aggregate at all show essentially the same patterns of protein synthesis as the wild-type for the first six hours after plating (Fig. 7B). However, these mutants continue making most of the "early" proteins, including actin, for at least twenty-four hours after plating. They show no induction of the polypeptides synthesized in wild-type cells during aggregation. As measured by the incorporation of [^{35}S]methionine into protein

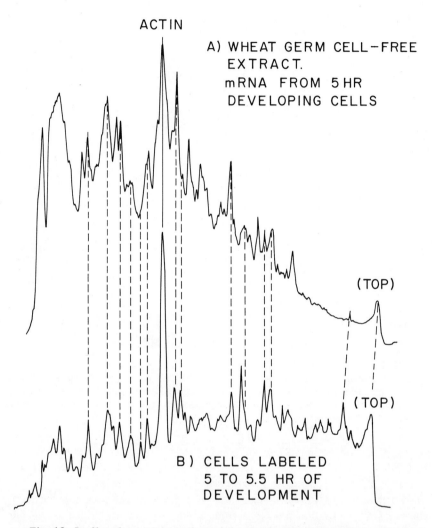

Fig. 10. *Profile of proteins synthesized by developing dictyostelium cells and in wheat germ extracts programmed with slime mold RNA. From whole* Dictyostelium *cells plated for 5 hr of development, RNA was isolated by lysis with phenol-chloroform-SDS, followed by precipitation with ethanol. The RNA was translated in the wheat germ extract as described by Roberts and Patterson (1973) and Mulligan and Roberts (manuscript in preparation) except that the reaction contained 800 µM spermidine and 1.5 mM Mg^{++}.*

Each 50 µl reaction contained 16 µg of whole-cell RNA. Whole cells were labeled from 5 to 5.5 hrs of development as described in Fig. 7A. The reaction products and in vivo proteins were subjected to electrophoresis on SDS-polyacrylamide gels containing an exponential gradient of acrylamide as described by Van Blerkom and Manes (1974). Autoradiograms were scanned with a Joyce-Loebl microdensitometer.

during a 30 minute label, the rate of protein synthesis by these cells declines by no more than two-fold during the first 24 hours (data not shown). We conclude that both the reduction in synthesis of certain early proteins and the increase in synthesis of late proteins after aggregation required gene function during differentiation. We do not know how these genes affect protein synthesis. Nor do we know at what level this control is exerted. One possibility is that cell-cell contact, or some excreted cell product which accumulates during aggregation, is responsible for induction of synthesis of new "aggregation-specific" mRNAs and for cessation of synthesis of "early" mRNAs. Whether, in addition, mRNA stability is developmentally controlled also remains for future work.

It is of interest that mutants which show some aggregation, but which do not form proper pseudoplasmodia, do show increases in synthesis of some, but not all, of the "aggregation-stage" proteins (Fig. 7C). However, these mutants, in contrast to wild-type cells, continue to synthesize these proteins for up to 24 hrs after initiation of development.

IV. MESSENGER RNA IN PREAGGREGATING CELLS

It is important to recall that in *Dictyostelium* differentiation is induced by starvation. Recent studies showed that starvation of amino acids — more precisely, of at least one of several essential amino acids — is the main trigger (F. Rothman, personal communication). Differentiation is fueled by endogenous energy reserves, and, as noted above, there is extensive turnover of both RNA and protein (reviewed, Jacobson and Lodish, 1975).

The earliest observed effect of starvation is an immediate reduction in the fraction of ribosomes in polyribosomes (Fig. 13A). In growing cells, over 90% of the cytoplasmic mRNA (Fig. 14) and 90% of the ribosomes (Fig. 13A) are found on polysomes; the average mRNA contains 8 to 12 ribosomes (Fig. 14). After as little as 5 minutes of development, about 60% of the ribosomes are present as inactive monosomes, and the few remaining polysomes are smaller than those in vegetative cells (Fig. 13B). The experiments in Figure 13 involve co-extraction of developing and vegetative cells and show that the polysomes are not degraded during extraction. As an additional control, if cells are plated on filters containing growth medium rather than buffer, the polysomes are essentially the same as in vegetative cells (Fig. 15). Additionally, if the cells are starved, but in liquid suspension rather than in filters, the reduction in polysome size and amount still occurs (Fig. 15). These results imply that there is an approximately three-fold reduction in the rate of initiation of protein synthesis.

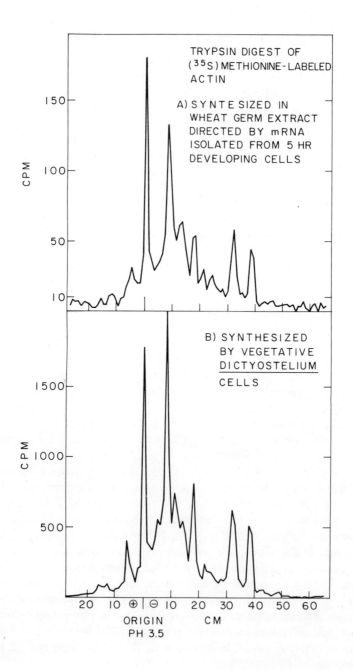

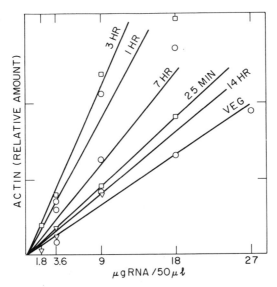

Fig. 12. *Relative amount of actin mRNA present in developing cells. RNA was prepared from whole cells, at stages indicated, as described in Fig. 10 and translated in the wheat germ extract. Products were analyzed on SDS-polyacrylamide gels which were fixed, dried and autoradiographed (see Fig. 7 and 10). The autoradiograms were then scanned and the area under the actin peak determined. Plotted is the area under the actin peak relative to the amount of input RNA. The slope of these curves are a measure of the content of actin mRNA.*

What causes this marked transition in the rate of protein synthesis? Several studies suggest that the complement of mRNA present in growing cells does not change significantly during early development. The amount of cytoplasmic RNA which can hybridize with labeled poly(U) — i.e., poly(A) in mRNA — does not change, but in preaggregating cells, over 25% of the mRNA is unattached to polysomes, and the remainder is on smaller polysomes than in growing cells (Fig. 14B). This implies that most of the vegetative mRNA is translated poorly in very early preaggregating cells.

Fig. 11. *(at left) Tryptic peptides of actin synthesized in vivo and in wheat germ extracts programmed with slime mold RNA. RNA from whole cells plated 5 hrs was prepared and translated in wheat germ extract as described in the legend to Fig. 10. Translation products and in vivo, synthesized proteins were separated on SDS-polyacrylamide gels which were fixed, dried down and subjected to autoradiography. The heavily labeled bands (actin) were excised and digested with trypsin as described by Tuchman, Alton and Lodish (1974). The resulting tryptic peptides were then subjected to high voltage paper ionophoresis at pH 3.5 (40 V/cm, 3 hr), the dried paper cut into 1 cm strips, and radioactivity determined by scintillation counting in a Toluene-POPOP cocktail.*

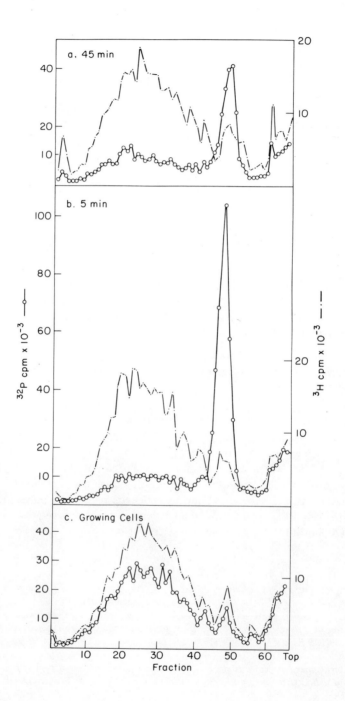

The experiment in Figure 16 suggests that preaggregating cells contain inactive mRNA which is intact and is potentially able to function efficiently as message. Cells which had been differentiating for 75 minutes, and which contained few polysomes, were removed from the filters and resuspended in growth medium. Within 15 minutes all of the ribosomes were found on normal-sized polysomes. Significantly, reformation of polysomes occurred almost normally in the presence of sufficient actinomycin D and daunomycin to completely suppress mRNA (and rRNA) biogenesis (Firtel, et al., 1973). This suggests that both the mRNA and ribosomes in preaggregating cells, while poorly active in protein synthesis, are potentially capable of functioning quite efficiently providing the cells are returned to a nutrient medium.

It is not clear to what extent this rapid drop in the rate of initiation of mRNA translation at the beginning of differentiation contributes to the changing pattern of protein synthesis. We know from work on other systems that different mRNAs have different rate constants for attachment to ribosomes and initiation of protein synthesis (Lodish, 1974). A similar situation may obtain for *Dictyostelium*. We also know that under conditions where the rate of polypeptide chain initiation is reduced non-specifically (say, by reduction in the number of active ribosome subunits or in the level of the initiator met-tRNA$_f$) translation of the "better" mRNAs — those with higher rate constants for chain initiation — will be inhibited less than that of "poorer" mRNAs. If one assumes that the rate constant for attachment of actin mRNA to the small ribosome subunit in the initiation step of protein synthesis is much greater than that of most other cellular RNAs, then the non-specific reduction in the rate of chain initiation during early development could account for at least part of the *relative* increase in the rate of actin synthesis.

Fig. 13. *(at left) Polysomes in differentiating Dictyostelium cells. Vegetative cells were labeled with either [^{32}P] (40 μCi/ml) or [^{3}H] uracil (70 μCi/ml) for 7 generations. Cells labeled with [^{32}P] were plated for differentiation; at the time indicated in panels a and b, they were mixed with [^{3}H]-labeled cells in ratio of 2:3. Polysomes were prepared essentially as described by Cocucci and Sussman (1970). Cells were lysed by rapid agitation at 0°C for 1 min in a buffer containing 50 mM Hepes pH 7.5, 40 mM MgAc$_2$, 20 mM CKl, 1 mg/ml Heparin, 5% sucrose and 4% Cemulsol NP 12 (Firtel and Lodish, 1973). Cellular debris was removed by rapid high speed centrifugation and the resulting cytoplasmic extract layered onto linear gradients of sucrose from 7% to 50% (w/v) made up in the above buffer, but without the Cemulsol and Heparin. The gradients were centrifuged at 26,500 rpm at 1°C for 3 hrs in a Beckman SW27 rotor. Gradients were collected and fractionated from the bottom; aliquots were precipitated in 6% ice cold TCA, filtered, dried and counted. (a) [^{32}P] vegetative cells mixed with [^{3}H] vegetative cells. (b) [^{32}P] cells after 5 min of differentiation mixed with [^{3}H] vegetative cells. (c) [^{32}P] cells plated 45 min mixed with [^{3}H] vegetative cells.*

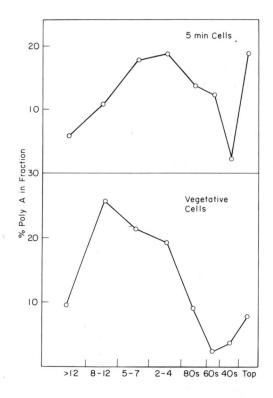

Fig. 14. *Messenger RNA in Dictyostelium cells. Cytoplasmic extracts from vegetative cells and cells 5 min into differentiation were prepared and run on sucrose gradients as described in Fig. 13. Fractions were pooled as shown, and RNA extracted by the SDS-phenol-chloroform method (Firtel and Lodish, 1973). The aqueous layers were precipitated with 70% ethanol in 0.2 M sodium acetate, pH 7.5 and the RNA precipitate recovered by centrifugation. The pellets were dissolved in sterile distilled water and hybridized to [^{3}H]-poly(U) (Miles Laboratories, Elkart, Ind.) by the method of Bishop, et al. (1974). After 15 min at 45°C in 2 X SSC, the 50 μl reactions were chilled, diluted 10-fold with ice cold SSC; then 20 μg/ml RNase A was added. The reactions were digested for 60 min at 25°C, chilled, precipitated with equal volumes of 12% ice cold TCA, filtered, dried and counted. Control experiments not shown here showed that the total amount of cytoplasmic polyadenylic acid is the same in these two preparations; the results shown here are expressed as percentage of total cytoplasmic poly(A) in each fraction of the polysome gradient. The numbers on the ordinate show the number or type of ribosome in each fraction. (a) Hybridization to RNA from vegetative cells. (b) Hybridization to RNA from cells at 5 min of differentiation.*

V. CODA

It is clear that further progress requires exact measurement of the absolute rates of synthesis, per cell, of several developmentally significant proteins, such as actin. Also, it is essential to purify mRNAs for many of these proteins, and also complementary DNA or RNA to use as hybridization probes for these mRNAs. Only then will it be possible to quantitate the rates of synthesis and destruction of specific mRNAs during differentiation. We will have to analyze protein and mRNA synthesis in a much larger number of aggregation-defective mutants, particularly mutants which are known to be in different complementation groups.

Finally, it is worth emphasizing that detailed understanding of even one agg $^-$ mutant will require the coordinated efforts of many laboratories — research groups whose primary interest is in *Dictyostelium* genetics, cAMP metabolism and binding, membrane changes, RNA and protein synthesis, and morphogenetic movements. The problem of differentiation in even this relatively simple system is far too complex to be attacked by a single laboratory.

ACKNOWLEDGEMENTS

We thank Ms. Lolley Gee and Mr. Martin Brock for expert technical assistance. This work was supported by Grants GB-42597 and BMS74-04869 from the National Science Foundation. H.F.L. is a recipient of a Research Career Development Award GM50175 from the U.S. National Institutes of Health; T.A. and J.P.M. have predoctoral fellowships from the N.S.F.; R.P.D. is a Centennial Fellow of the Canadian Medical Research Council; and A.M.W. is a Helen Hay Whitney postdoctoral fellow.

REFERENCES

Adams, J. and Cory, S. (1975). *Nature* **255**, 28–33.
Ashworth, J. M. (1971). *Symp. Soc. Exp. Biol.* **25**, 27–49.
Bishop, J. W., Rosbash, M. and Evans, D. (1974). *J. Mol. Biol.* **85**, 75–88.
Bonner, J. T. (1971). *Ann. Rev. Microbiol.* 75–92.
Bonner, J. T., Clarke, W. W., Jr., Neely, C. L., Jr. and Slifkin, M. K. (1950). *J. Cell Comp. Physiol.* **36**, 149–158.
Clark, M. A., Francis, D., and Eisenberg, R. (1973). *Biochem. Biophys. Res. Comm.* **52**, 672–678.
Cocucci, S. M. and Sussman, M. (1970). *J. Cell Biol.* **45**, 399–407.
Darnell, J. E., Jelinek, W. R. and Molloy, G. R. (1973). *Science* **181**, 1215–1218.
Durston, A. (1974). *Develop. Biol.* **37**, 225–235.
Erdos, G. W., Raper, K. B. and Vogen, L. K. (1973). *Proc. Nat. Acad. Sci. U.S.A.* **70**, 1828–1830.
Erdos, G. W., Raper, K. B. and Vogen, L. K. (1975). *Proc. Nat. Acad. Sci. U.S.A.*, in press.

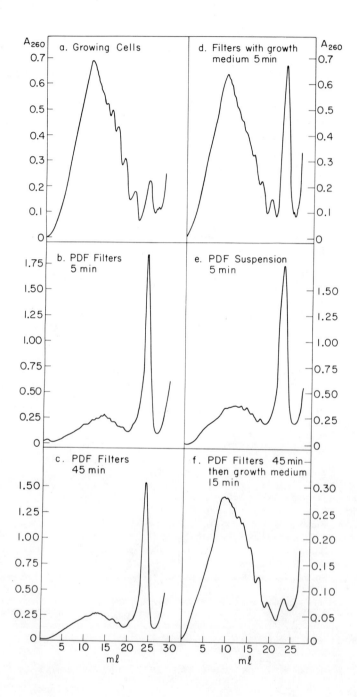

Firtel, R. A. (1972). *J. Mol. Biol.* **66**, 363–378.
Firtel, R. A., Baxter, L. and Lodish, H. F. (1973). *J. Mol. Biol.* **79**, 315–327.
Firtel, R. A. and Bonner, J. (1972). *J. Mol. Biol.* **66**, 339–362.
Firtel, R. and Kindle, K. (1975). *Cell*, in press.
Firtel, R. A., Jacobson, A. and Lodish, H. F. (1972). *Nature New Biol.* **239**, 225–228.
Firtel, R. and Lodish, H. F. (1973). *J. Mol. Biol.* **79**, 295–314.
Franke, J. and Sussman, M. (1973). *J. Mol. Biol.* **81**, 173–185.
Furuichi, Y., Morgan, M., Mothukrishnan, S. and Shatkin, A. (1975a). *Proc. Nat. Acad. Sci. U.S.A.* **72**, 362–366.
Furuichi, Y., Morgan, M., Shatkin, A. J., Jelinek, W., Salditt-Georgieff, M. and Darnell, J. E. (1975b). *Proc. Nat. Acad. Sci. U.S.A.* **72**, 1904–1908.
Gerisch, G., Malchow, D. and Hess, B. (1974). *In* "Biochemistry of Sensory Functions" (L. Jaenicke, ed.), pp. 279–298. Springer-Verlag, Berlin.
Gingold, E. B. and Ashworth, J. M. (1974). *J. Gen. Microbiol.* **84**, 70–78.
Hohl, H. R. and Hammamoto, S. T. (1969). *J. Ultrastruct. Res.* **26**, 442–453.
Jacobson, A., Firtel, R. A. and Lodish, H. F. (1974a). *J. Mol. Biol.* **82**, 213–230.
Jacobson, A., Firtel, R. A. and Lodish, H. F. (1974b). *Proc. Nat. Acad. Sci. U.S.A.* **71**, 1607–1611.
Jacobson, A. and Lodish, H. F. (1975). *Ann. Rev. Genetics,* **Vol. 9**, in press.
Katz, E. and Bourguignon, L. (1974). *Develop. Biol.* **36**, 82–87.
Katz, E. R. and Kao, V. (1974). *Proc. Nat. Acad. Sci. U.S.A.* **71**, 4025–4026.
Kessin, R. H., Williams, K. L. and Newell, P. C. (1974). *J. Bacteriol.* **119**, 776–783.
Killick, K. A. and Wright, B. E. (1974). *Ann. Rev. Microbiol.* **28**, 139–166.
Laemmli, O. (1970). *Nature* **227**, 680–684.
Lodish, H. F. (1974). *Nature* **251**, 385–388.
Lodish, H. F., Firtel, R. and Jacobson, A. (1973). *Cold Spring Harbor Symp. Quant. Biol.* **38**, 899–914.
Lodish, H. F., Jacobson, A., Firtel, R., Alton, T. and Tuchman, J. (1974). *Proc. Nat. Acad. Sci. U.S.A.* **71**, 5103–5108.
Loomis, W. F., Jr. (1969). *J. Bacteriol.* **99**, 65–69.
Loomis, W. F., Jr. (1975). "Dictyostelium discoideum, a Developmental System". Academic Press, New York, in press.
Milcarek, C., Price, R. and Penman, S. (1974). *Cell* **3**, 1–10.
Nakazato, H., Edmonds, M. and Kopp, D. W. (1974). *Proc. Nat. Acad. Sci. U.S.A.* **71**, 200–204.
Newell, P. C. (1971). *In* "Essays in Biochemistry" (P. N. Campbell and F. Dickens, eds.), Vol. 7, pp. 87–126. Academic Press, New York and London.
Perry, R. P., Kelley, D. E., Friderici, K. and Rottman, F. (1975). *Cell* **4**, 387–394.
Raper, K. B. (1940). *J. Elisha Mitchell Sci. Soc.* **56**, 241–282.
Roberts, B. and Paterson, B. (1973). *Proc. Nat. Acad. Sci. U.S.A.* **70**, 2330–2334.
Robertson, A. and Cohen, M. H. (1972). *Ann. Rev. Biophys. Bioengineering* **1**, 409–464.

Fig. 15. *(at left) Profiles of polysomes in Dictyostelium cells. Polysomes were prepared and analyzed as described in Fig. 13; the gradients were pumped through a 5 mm flow cell and the absorbance at 260 nm monitored continuously. Sedimentation is from right to left. (a) Growing axenic cells. (b) Cells plated for 5 min on Millipore filters in PDF (buffered saline). (c) Cells plated for development for 45 min as in (b). (d) Cells plated on Millipore filters containing growth medium. (e) Cells shaken in PDF suspension for 5 min. (f) Cells plated for development 45 min [as in (c)] then returned to growth medium for 15 min.*

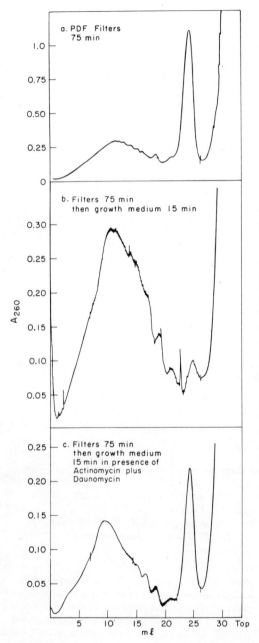

Fig. 16. *Reversibility of polysome profiles. After 75 min of differentiation, cells were washed from the filters and resuspended at 2 X 10^6 cells/ml in growth medium some (panel c) containing 125 µg/ml actinomycin D plus 250 µg/ml daunomycin (Firtel, et al., 1973), and some (panel b) without. The resuspended cells were aerated 15 min under standard growth conditions, then recovered by centrifugation. Cytoplasmic extracts were prepared and run on sucrose gradients as described in Fig. 13. (a) Cells at 75 min of differentiation. (b) Cells at 75 min of differentiation which were then returned to growth medium 15 min. (c) As in (b), except that the growth medium contained 125 µg actinomycin D and 250 µg daunomycin per ml.*

Sussman, M. (1966). *In* "Methods in Cell Physiology" (D. Prescott, ed.), pp. 397–410. Academic Press, New York.
Sussman, R. R. and Sussman, M. (1967). *Biochem. Biophys. Res. Comm.* **29,** 53–55.
Tuchman, J., Alton, T. and Lodish, H. F. (1974). *Develop. Biol.* **40,** 116–128.
Van Blerkom, J. and Manes, C. (1974). *Develop. Biol.* **40,** 40–51.
Watts, D. J. and Ashworth, J. M. (1970). *Biochem. J.* **119,** 171–174.
Wei, C. M., Gershowitz, A. and Moss, B. (1975). *Cell* **4,** 379–386.
Weinberg, R. A. (1973). *Ann. Rev. Biochem.* **42,** 329–354.
Williams, K. L., Kessin, R. H. and Newell, P. C. (1974). *J. Gen. Microbiol.* **84,** 59–69.

# II. Action of Steroid Hormones

REGULATION OF TRANSCRIPTION OF THE EUCARYOTIC GENOME

H. C. Towle, M.-J. Tsai, M. Hirose, S. Y. Tsai,
R. J. Schwartz, M. G. Parker and B. W. O'Malley

*Department of Cell Biology,
Baylor College of Medicine,
Houston, Texas 77025*

I. INTRODUCTION

In the past decade a great deal of progress has been made towards an understanding of the regulation of gene expression in procaryotic systems (for review see; Lewin, 1974). Much of this progress has been achieved by studying the *in vitro* transcription of well-defined bacterial and bacteriophage DNA templates with purified bacterial RNA polymerase. Through the use of these systems, the roles of many regulatory elements, such as specific gene repressors, activators, and termination factors, have been elucidated. A complete understanding of the positive and negative controls which function to regulate the transcription of certain bacterial operons, notably the lactose operon of *Escherichia coli*, seems close at hand. Furthermore, the ability to faithfully reconstitute the cellular transcription of the lactose operon by manipulation of the various purified components indicates the degree of sophistication which has been attained. With the tools of molecular biology and genetics currently available, studies on the regulation of other genetic elements of bacterial and bacteriophage-infected cells should continue to make substantial progress.

The current understanding of the regulation of gene expression in eucaryotic organisms lags behind work in the procaryotic field. Analogous studies on the *in vitro* transcription of eucaryotic genomes pose more difficult technical problems. For instance, the genetic complexity of eucaryotic organisms is several orders of magnitude greater than that of procaryotic organisms, making the detection of specific gene products a considerably more arduous task. While the use of genetic mutations has proven to be an immeasurable aid in defining the role of many regulatory elements in bacteria, few workable systems for obtaining genetic information exist for eucaryotes. Furthermore, the DNA of higher organisms is complexed with a tremendous variety of histone and nonhistone proteins in a structure which remains largely undefined. Even the nature of the primary transcription product leading to the synthesis of messenger RNA is still in question. Such problems have inhibited progress towards understanding the regulation of gene expression in eucaryotes.

Our approach to study gene regulation has been to utilize chromatin as a template for *in vitro* transcriptional studies. Since the histone and nonhistone chromosomal proteins may play an integral role in gene expression, we feel the proper template for *in vitro* transcriptional studies is chromatin and not deproteinized DNA. Our initial efforts have been to characterize the interaction of purified RNA polymerase with unsheared chromatin. The process of initiation of RNA synthesis on chromatin has received special attention, as this step appears to provide the greatest potential for controlling gene expression. In this article, we will discuss our recent findings on the transcription of chromatin isolated from chick oviduct.

The chick oviduct provides an excellent model system for the study of the control of gene expression. Chronic administration of estrogenic compounds to the immature chick over a period of 10 to 12 days results in the growth and differentiation of the oviduct (O'Malley et al., 1969). Several specific proteins are produced by the oviduct after estrogen administration. These changes in the oviduct appear to be mediated mainly at the level of transcription (O'Malley and Means, 1974). Dramatic changes in oviduct endogenous RNA polymerase activity (O'Malley et al, 1969; Cox et al., 1973), nuclear RNA synthesis (O'Malley and McGuire, 1968; Means et al., 1972; Harris et al., 1975a), and chromatin template activity (O'Malley, et al., 1969; Cox et al., 1973; Spelsberg et al., 1973) occur following estrogen administration. Furthermore, the appearance of messenger RNA for a specific induced protein, ovalbumin, precedes the accumulation of that protein during estrogen-mediated growth (Means et al., 1972; Comstock, et al., 1972; Rhoads et al., 1973). If the estrogen stimulated chick is withdrawn from steroid hormone, the mRNA for ovalbumin gradually disappears from the oviduct over a period of several days (Harris et al., 1975a). During withdrawal, however, the oviduct appears to remain in a quasi-differentiated state, as readministration of estrogen results in the production of ovalbumin mRNA within 1 hour (Chan et al., 1973; Cox et al., 1974; Palmiter, 1974; Harris et al., 1975a). Consequently, the chick oviduct provides an opportunity to study a specific eucaryotic tissue in which the expression of genetic information can be dramatically altered. Furthermore, a specific marker for following the change in genetic expression is available. We have, thus, undertaken a study of the *in vitro* transcription of chick oviduct chromatin by exogenously added RNA polymerase. From this study, we hope to learn the mechanism by which gene expression is altered following hormone uptake by estrogen responsive cells. In so doing, it is likely that much will be learned about the basic process of transcription in eucaryotic organisms.

II. KINETICS OF RNA CHAIN INITIATION

The initiation of RNA synthesis can be divided into two basic processes. The first involves selection of the proper site on DNA where initiation is to occur and formation of a binary complex between RNA polymerase and DNA at this site. The second process is the actual initiation of an RNA chain by formation of the first phosphodiester bond between two nucleoside triphosphates. As a first step towards a study of the *in vitro* transcription of chromatin, the kinetics of these two processes on chick oviduct chromatin and on deproteinized DNA were examined utilizing *E. coli* RNA polymerase.

Chamberlin and co-workers have extensively studied the interaction between *E. coli* RNA polymerase and bacteriophage T_7 DNA leading to the formation of binary complexes capable of initiating RNA synthesis (Chamberlin, 1974). Utilizing this system, the formation of binary complexes was shown to occur through several intermediate steps:

$$\underset{\text{Nonspecific complex}}{\overset{(1)}{(DNA\text{-}ENZ)}} \rightleftarrows DNA + ENZ \overset{(2)}{\rightleftarrows} \underset{\text{I complex}}{(DNA\text{-}ENZ)_I} \overset{(3)}{\rightleftarrows} \underset{\text{RS complex}}{(DNA\text{-}ENZ)_{RS}}$$

When RNA polymerase is incubated with DNA, the enzyme initially binds randomly and reversibly to DNA to form a series of nonspecific complexes (step 1). If such binding occurs at or near a true initiation site for RNA synthesis, an initial preinitiation complex (I complex) is formed (step 2). This complex must then undergo a transition, involving the local opening of the DNA duplex structure, to form a highly stable complex (RS complex) capable of rapidly initiating RNA synthesis (step 3).

To measure the formation of the highly stable RS complex, the bacterial RNA polymerase inhibitor, rifampicin, is utilized. Rifampicin is an inhibitor of RNA synthesis which acts prior to the formation of the first phosphodiester bond, but which has no effect on RNA chain elongation (Sippel and Hartman, 1968; Umezawa et al., 1968). When RNA polymerase present in the RS complex is challenged with a mixture of the four ribonucleoside triphosphates and rifampicin, the enzyme is highly efficient at initiating RNA synthesis. The lack of inhibition of this complex has been shown to be due to the rapid rate of RNA chain initiation, rather than the RS complex having an absolute resistance to rifampicin (Hinkle et al., 1972; Mangel and Chamberlin, 1974). Enzyme molecules which are bound at nonspecific or I complexes or are free in solution, however, are largely incapable of initiating RNA synthesis in such a challenge experiment. Thus, the use of the rifampicin-ribonucleotide challenge experiment allows the differentiation of RNA polymerase present at RS complex from other forms of the enzyme.

To follow the kinetics of formation of RS complexes, RNA polymerase is first preincubated with template. At various times, RNA synthesis is initiated by the simultaneous addition of rifampicin and ribonucleotide. The amount of RNA synthesis at each time point is directly proportional to the amount of RS complex formed. The results of such a time course of RS complex formation with either chick DNA or chick oviduct chromatin is shown in Figure 1. For this experiment, chromatin isolated from fully stimulated chicks (14 days of diethylstilbestrol [DES] treatment) was used. The formation of RS complexes on chick DNA reached a maximum level by about 10 minutes of preincubation. On chromatin, the maximum was not reached until much longer preincubation periods of about 40 minutes. As shown on the insets to Figure 1, semilogarithmic plots of the maximum level of RNA synthesis minus the level of RNA synthesis at each preincubation time versus the time of preincubation were linear for both templates. This result would only be expected if the formation of RS complexes occurred via a first order process. From these plots, the half time of formation of the RS complex ($t_{1/2}$) between $E.$ $coli$ RNA polymerase and chick DNA was found to be 1.3 minutes. On the other hand, the $t_{1/2}$ of formation of the RS complex on chromatin was 9.4 minutes or approximately seven times slower than on deproteinized DNA.

The slow rate of RS complex formation on chromatin might possibly be attributed to a gradual change in the chromatin structure during the preincubation period, leading to the exposure of new initiation sites. To test this possibility, chromatin was incubated in the absence of RNA polymerase for varying lengths of time. This pretreated chromatin was then preincubated with RNA polymerase for 15 minutes. The level of RNA polymerase activity (Figure 1B, open triangles) was constant for chromatin pretreated for periods up to 60 minutes and closely corresponded to the activity of chromatin which had received no pretreatment. Thus, it is unlikely that any artifactual modification of the chromatin structure occurs during 60 minutes of preincubation. The level of endogenous RNA polymerase activity in the chromatin was also checked by measuring RNA synthesis in the absence of added $E.$ $coli$ RNA polymerase. The endogenous RNA polymerase activity (closed triangles) did not contribute significantly to the total level of RNA synthesis and thus the $t_{1/2}$ measured actually represents the formation of RS complexes between $E.$ $coli$ RNA polymerase and chromatin. Since this $t_{1/2}$ is dramatically slower than that for deproteinized chick DNA, chromosomal proteins must somehow interact with RNA polymerase to retard the rate of RS complex formation. The probable nature of this interaction will be discussed in a later section.

The actual initiation of RNA chain synthesis, involving the formation of a 5′ to 3′ phosphodiester linkage between two nucleoside triphosphates, occurs directly from the RS complex. By the use of the ribonucleotide-rifampicin challenge technique, the rate constant of RNA chain initiation can be measured. As demonstrated by Mangel and Chamberlin (1974), the formation of the first phosphodiester bond in the RNA chain and the attack of rifampicin are competing reactions for any available RS complexes. The rate constant of RNA chain initiation from preformed RS complexes can be determined by measuring RNA polymerase activity in the presence of varying concentrations of rifampicin in such a challenge experiment. The theoretical basis for this determination is expressed in the equation:

$$\frac{C_0}{C^*} = \frac{k_2 [R]}{k^*} + 1 \qquad (1)$$

where C_0 represents the total concentration of RS complex present at the time of addition of ribonucleotide and rifampicin; C^* equals the concentration of RS complexes which are able to initiate RNA synthesis at a given concentration of rifampicin, R; k_2 represents the second order rate constant of rifampicin attack on RS complexes; and k^* is the apparent first order rate constant for RNA chain initiation at a fixed concentration of ribonucleotide (Mangel and Chamberlin, 1974). In the absence of any secondary initiations from a single site, the ratio of C_0 to C^* should be proportional to the ratio of RNA polymerase activity in the absence of rifampicin (V_0) to that in the presence of rifampicin (V^*). Thus, from the slope of the plot of V_0/V^* versus the concentration of rifampicin, the value of k_2/k^* can be obtained. The second order rate constants of rifampicin attack on RS complexes between *E. coli* RNA polymerase and either chick DNA or chick oviduct chromatin have been measured (Hirose et al., 1975). The k_2 values for chick DNA and chromatin are 2.3×10^3 sec^{-1} M^{-1} and 0.8×10^3 sec^{-1} M^{-1}, respectively. Therefore, it should be possible to obtain the rate constant for RNA chain initiation from RS complexes for these two templates from a plot of V_0/V^* versus the concentration of rifampicin.

The results of such a determination utilizing either chick DNA or chick oviduct chromatin are shown in Figure 2. For this experiment, RNA polymerase and template were first preincubated to allow maximum formation of RS complexes. RNA synthesis was initiated by the addition of 0.15 mM each of the four ribonucleoside triphosphates and varying concentrations of rifampicin, as indicated. The incubation period for RNA synthesis was limited to 90 seconds to minimize any secondary initiation events in the absence of rifampicin. As shown on the inset, a plot of V_0/V^* versus rifampicin concentration was linear on both DNA and chromatin, as predicted by equation (1). The intercepts on the ordinates for these plots, however, which should theoretically be 1.0, were 1.4 for DNA and 4.0 for chromatin. The most likely explanation for this difference is that in the absence of rifampicin secondary initiation events from a single site occur even within 90 seconds. Such reinitiations could not occur in the presence of rifampicin. Therefore, the value of V_0 is artifactually high and must be corrected by a factor equal

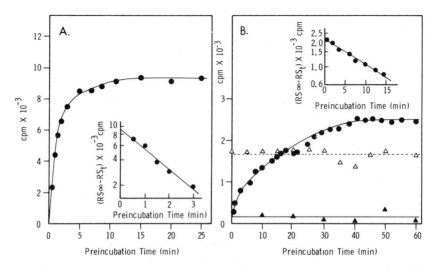

Fig. 1. *Time course of formation of RS complex. (A)* E. coli *RNA polymerase (1.4 μg) was incubated at 37°C with chick DNA (0.75 μg) in 0.1 ml of preincubation buffer containing 62.5 mM Tris-HCl, pH 7.9, 1.25 mM $MnCl_2$, 62.5 mM $(NH_4)_2SO_4$, 2.5 mM 2-mercaptoethanol, and 0.5 mg/ml bovine serum albumin. After the indicated time interval, 0.025 ml of ribonucleoside triphosphate mixture containing 0.75 mM each of ATP, CTP, GTP, [^3H]UTP (0.2 mC/ml), and 0.2 mg/ml rifampicin was added. RNA synthesis was carried out at 37°C for 1.5 minutes and reactions terminated by the addition of 5 ml of cold 5% trichloroacetic acid containing 0.01 M sodium pyrophosphate. Samples were analyzed for incorporation of [^3H]UTP into acid-insoluble material as described elsewhere (Hirose et al., 1975). (B)* E. coli *RNA polymerase (7.0 μg) was incubated at 37°C with chick oviduct chromatin (5.0 μg) in 0.2 ml of preincubation buffer for the indicated time intervals (●). Controls were run in which 5.0 μg of chromatin was incubated without any added enzyme (▲) to test for endogenous RNA polymerase activity and in which 5.0 μg of chromatin was preincubated for the indicated time interval without any added enzyme, followed by the addition of 7.0 μg of RNA polymerase for 15 minutes (△) to test for changes in chromatin during incubation. RNA synthesis was initiated by the addition of 0.05 ml of ribonucleoside triphosphate mixture. Chromatin was isolated by the method of Tsai et al. (1975) from the oviducts of chicks which had received daily injections of diethylstilbestrol (2.5 mg) for 14 days (14 day DES stimulated). All other details were as described above.*

to the reciprocal of the intercept. When this correction is made, the apparent k* values obtained were 0.98 sec^{-1} for chick DNA and 0.57 sec^{-1} for chromatin. These values correspond to a $t_{1/2}$ for RNA chain initiation of 0.7 seconds for DNA and 1.2 seconds for chromatin. Thus, the rate of RNA chain initiation from RS complexes was much faster than the rate of formation of such complexes. Furthermore, the k* values obtained for chick DNA and chick oviduct chromatin were not remarkably different. The rate of formation of RS complexes on chromatin, however, was seven times slower than that for DNA.

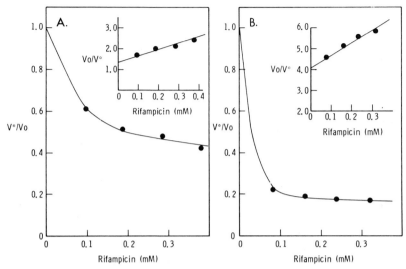

Fig. 2. *Determination of the rate constants of RNA chain initiation. (A)* E. coli *RNA polymerase (2.0 μg) and chick DNA (1.5 μg) were incubated for 15 minutes at $37°C$ in 0.2 ml of preincubation buffer to allow formation of RS complex. RNA synthesis was initiated by the addition of 0.05 ml of ribonucleoside triphosphate mixture in which the concentration of rifampicin was varied to yield the final concentrations indicated. V^*, the level of [^{3}H]UTP incorporation in the absence of rifampicin, was equal to 22,500 CPM. Other conditions were as described in the legend to Figure 1. (B) RS complex was formed by preincubating 7.0 μg of* E. coli *RNA polymerase with 5.0 μg of chick oviduct chromatin (14 day DES stimulated) in 0.2 ml of preincubation buffer for 40 minutes. V^* was equal to 2700 CPM of [^{3}H]UTP incorporation. Other conditions were the same as above.*

III. DETERMINATION OF THE NUMBER OF INITIATION SITES ON CHROMATIN

The use of the ribonucleotide-rifampicin challenge technique provides a method for estimating the number of available initiation sites for RNA polymerase in a given template (Bautz and Bautz, 1970; Chamberlin and Ring, 1972; Tsai et al., 1975). The method involves first preincubating RNA polymerase and template for a period of time sufficient to allow maximum formation of RS complex. RNA synthesis is then initiated by the addition of the four ribonucleoside triphosphates with 40 μg/ml rifampicin and allowed to proceed to completion. At this concentration of rifampicin, initiation from the RS complex is highly efficient, while secondary initiations are strongly inhibited (Mangel and Chamberlain, 1974; Tsai et al., 1975; Hirose et al., 1975). Thus, the number of RNA chains synthesized, which can be determined from the amount, length, and base composition of the RNA produced, will be equivalent to the number of initiation sites at which RNA polymerase is bound. By adding increasing amounts of RNA polymerase to a fixed amount of template, the total number of available initiation sites can be determined from the saturation level of RNA synthesis obtained when all high affinity binding sites are occupied.

The results of such an analysis for chick DNA is shown in Figure 3. Increasing amounts of E. coli RNA polymerase holoenzyme or core enzyme (without sigma factor) were preincubated for 15 minutes at either 0° or 37°. Following preincubation, RNA synthesis was carried out for 15 minutes at 37°. The number of initiation sites available for holoenzyme after preincubation at 37° (curve A) was calculated to be about 1.3×10^6 per picogram of DNA (based on a number average chain length of 450 nucleotides and a base composition of 40.6% UMP). This corresponds to an average of one initiation site per 700 nucleotide base pairs of DNA. After preincubation at 0°, neither holoenzyme nor core enzyme (curve C, D) is very efficient at formation of RS complexes. We attribute this to the high temperature dependency of the conversion from I to RS complex. When core enzyme is preincubated at 37° (curve B), the number of RNA chains initiated is still only 30% of that seen

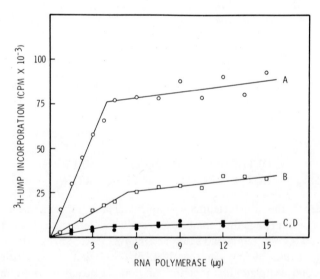

Fig. 3. *RNA initiation sites on chick oviduct DNA.* E. coli *RNA polymerase holoenzyme (○,●) or core enzyme (□,■) were incubated with 1.5 μg of chick oviduct DNA in 0.2 ml of preincubation buffer. After incubation for 15 minutes at either 37°C (○,□) or 0°C (●,■), RNA synthesis was initiated by the addition of 0.05 ml of ribonucleoside triphosphate mixture containing 0.2 mg/ml of rifampicin. RNA synthesis was carried out at 37°C for 15 minutes. Core enzyme was prepared by phosphocellulose chromatography of purified holoenzyme as described by Burgess and Travers (1971).*

for holoenzyme preincubated at 37°. Thus, both sigma factor and preincubation at 37° are required for the most efficient formation of RS complexes between E. coli RNA polymerase and chick DNA.

The number of initiation sites in chromatin isolated from a fully stimulated chick oviduct was analyzed in an analogous manner (Figure 4). In this experiment, chromatin (5 µg) was preincubated for 40 minutes with RNA polymerase. RNA synthesis was carried out in the presence of rifampicin and heparin. Heparin is a polyanion added to inhibit any RNase activity which, if present, would decrease the average length of the RNA produced. As the amount of RNA polymerase was increased, a transition point was observed which corresponded to the number of available high affinity binding sites in chromatin. The level of RNA synthesis at this point was used to calculate the number of sites available for the initiation of transcription. The number of initiation sites available for holoenzyme after preincubation at 37° was calculated to be about 22,500 per picogram of DNA (based on a number average chain length of 750 nucleotides and a base composition of 37.9% UMP). This corresponds to an average of one initiation site for every 40,000 base pairs of DNA. Thus, chromatin from the estrogen-stimulated chick

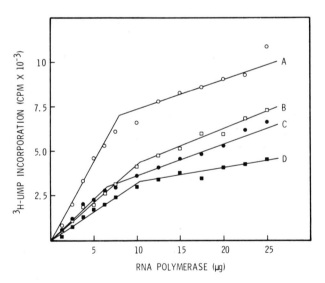

Fig. 4. *RNA initiation sites on chick oviduct chromatin.* E. coli *RNA polymerase holoenzyme (○,●) or core enzyme (□,■) were incubated with 5.0 µg of chick oviduct chromatin (14 day DES stimulated) in 0.2 ml of preincubation buffer. After incubation for 40 minutes at either 37°C (○,□) or 0°C (●,■), RNA synthesis was initiated by the addition of 0.05 ml of ribonucleoside triphosphate mixture containing 0.2 mg/ml rifampicin and 4.0 mg/ml heparin. Other conditions were as described in the legend to Figure 3.*

oviduct contained only about 2% of the total initiation sites as deproteinized DNA per picogram of DNA. Furthermore, the formation of RS complexes on chromatin was much less dependent on either the presence of sigma factor or the temperature of preincubation than that for DNA. Therefore, the initiation sites available for RNA polymerase in chromatin may be in a state in which the conversion from I to RS complex is greatly facilitated. One possible explanation is that certain chromosomal proteins are present at initiation sites which help to destabilize the duplex structure of the DNA.

The capacity of oviduct chromatin to serve as a template for *E. coli* RNA polymerase has previously been shown to increase following estrogen administration to the immature chick (Cox *et al.*, 1973; Spelsberg *et al.*, 1973). Such measurements, however, do not distinguish between the many steps of the RNA synthetic process which could influence determinations of total template capacity. We have, therefore, separately measured the number of high affinity initiation sites, the rate of RNA chain propagation, and the chain length of RNA synthesized from chromatin isolated from various stages during the primary stimulation of chick oviduct (Schwartz *et al.*, 1975). The number of initiation sites available for *E. coli* RNA polymerase was shown to increase by 50% as early as 8 hours after estrogen administration (Table 1). The increase in initiation sites reached a maximum by about 4 days of estrogen treatment and then declined to a plateau level, approximately two times higher than untreated animals, by 10 days. By this time estrogen-mediated differentiation is nearing completion in the chick oviduct. Over the same period the average rate of RNA chain elongation for the first minute of RNA synthesis and the average chain length of the RNA product did not vary significantly. Thus, the estrogen-induced increase in total chromatin template capacity is mainly due to an increased availability of sites at which RNA polymerase can bind and form RS complexes.

During primary estrogen stimulation of the chick oviduct, differentiation of three distinct epithelial cell types occurs from a homogeneous population of primitive mucosal cells (O'Malley *et al.*, 1969). Thus, changes in the number of initiation sites in oviduct chromatin during primary stimulation might simply reflect alterations in the tissue population of specific cell types. To overcome this difficulty, the number of available initiation sites for RNA polymerase was measured during estrogen withdrawal and acute (secondary) estrogen stimulation (Tsai, S.Y. *et al.*, 1975). During secondary stimulation, estrogen-mediated changes in RNA and protein syntheses are not dependent on cell proliferation (Palmiter *et al.*, 1970). As shown in Table II, the level of available initiation sites for *E. coli* RNA polymerase decreases after withdrawal of fully stimulated chicks to a level similar to that observed for unstimulated chicks. As early as 30 minutes after injection of a single dose of diethylstilbestrol, the number of initiation sites increased from 8100 to 17150. By 1 to 2 hours, a plateau level had been reached which was about 3 times higher than

TABLE I.
*Effect of Primary Estrogen Stimulation on
Initiation Sites for RNA Synthesis*[a]

Days of Hormone Treatment	Size of RNA Product[b]	Rate of Elongation (nucleotides/sec)	pmol of RNA Chains per 5 μg Chromatin	Initiation Sites per pg of DNA
0	660	6.0	0.09	10600
0.3	700	6.0	0.13	16800
1	810	6.0	0.28	33450
4	640	6.0	0.60	72000
8	725	---	0.35	42200
12	680	---	0.17	21200
18	700	5.8	0.23	28600

[a] Oviduct chromatin was isolated from chicks which had received daily injections of 2.5 mg of DES for the indicated periods. Determination of the size of the RNA product, the rate of RNA chain elongation, and the number of RNA initiation sites per pg of DNA were described elsewhere (Schwartz et al., 1975).

[b] Number average chain length of RNA product in nucleotides.

TABLE II.
*Effect of Estrogen Withdrawal and Secondary Estrogen Stimulation on
Initiation Sites for RNA Sysnthesis*[a]

Hormone Treatment	Size of RNA Product[b]	Rate of Elongation (nucleotide/sec)	pmol of RNA Chains/5 μg Chromatin	Initiation Sites per pg of DNA
18 day DES Stimulated	700	6.0	0.222	26700
18 day DES Stimulated + 12 day withdrawn	873	8.5	0.067	8100
18 day DES Stimulated + 12 day withdrawn + Secondary DES Stimulation for:				
20 min	810	7.0	0.082	9800
30 min	750	7.7	0.142	17100
60 min	621	8.0	0.205	24800
2 hr	777	8.0	0.223	26900
4 hr	803	7.0	0.212	25700

[a] Oviduct chromatin was isolated from chicks which were treated with hormone as described in the legend to Figure 5. Determination of the size of the RNA product, the rate of RNA chain elongation, and the number of RNA initiation sites per pg of DNA were described elsewhere (Tsai, S.Y. et al., 1975).

[b] Number average chain length of RNA product in nucleotides.

the withdrawn level. Measurements of the average chain length of the RNA product and the rate of RNA chain elongation were similar for all preparations tested. Thus, the change in chromatin template activity during secondary stimulation actually reflects an increase in the availability of initiation sites for RNA polymerase.

During secondary stimulation with estrogen, simultaneous measurements were also made of the level of endogenous nuclear receptor and the amount of intracellular ovalbumin mRNA. The level of nuclear bound receptor, measured by a [^{3}H] estradiol exchange assay (Anderson et al., 1972), increased to a transient maximum shortly after 20 minutes of readministration and then declined to a plateau level at 1 to 2 hours (Figure 5). The increase in the concentration of nuclear receptor molecules immediately preceded the increase in available initiation sites on chromatin prepared from the same nuclei. The measurement of the intracellular concentration of ovalbumin mRNA was made using a [^{3}H] labeled complementary DNA probe (Harris et al., 1973). A detectable increase in ovalbumin mRNA occurred at 30 to 60 minutes, followed by a linear accumulation beginning after 1 hour. The increase in the intracellular level of specific induced mRNA, thus, occurs shortly after the increase in the number of available chromatin initiation sites. The interesting temporal relationship of the three measured parameters makes it tempting to speculate a causal relationship between the events. It is possible that after translocation to the nucleus, hormone-receptor complexes may act directly on chromatin such that previously repressed DNA sequences on chromatin become accessible for transcription by RNA polymerase.

IV. COMPETITION OF *E. coli* AND HEN OVIDUCT RNA POLYMERASES FOR INITIATION SITES ON CHROMATIN

The nature of the initiation process on chromatin differed from that on deproteinized DNA. The formation of RS complexes between *E. coli* RNA polymerase and chromatin was much less dependent on the temperature of preincubation or the presence of sigma factor than the formation of such complexes on DNA. This suggested that certain structural features of the chromatin, perhaps specific chromosomal proteins, facilitated the conversion from I (closed form) to RS (open form) complex. If this is true, then such structural features would be expected to be major determinants of the location of initiation sites in chromatin. To test this possibility, we have attempted to learn whether different RNA polymerases utilize common or unique sites for initiation of RNA synthesis. The RNA polymerases examined were *E. coli* RNA polymerase holoenzyme, core enzyme, and hen oviduct RNA polymerase II. If these RNA polymerases utilize the same initiation sites on chromatin, despite having unique specificities on native DNA, it would

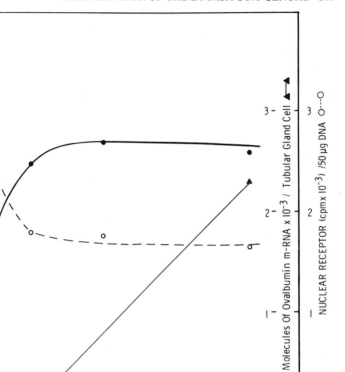

Fig. 5. *Effect of secondary estrogen stimulation on the initiation of RNA synthesis and the level of nuclear receptors for estrogen. Chicks which had received daily injections of 2.5 mg of DES for 18 days were withdrawn from hormone treatment for 12 days. Chicks were then given a single 2.5 mg dose of DES and at the indicated times, oviducts were removed and frozen. Chromatin was prepared from a portion of each oviduct sample for the measurement of initiation sites as described in text and in Table II. On the other portion of oviduct, the number of molecules of ovalbumin mRNA per tubular gland cell and the level of nuclear estrogen receptor were measured as described in the text.*

seem likely that elements in the chromatin structure are playing a role in the determination of initiation sites.

The previous results on the determination of the number of initiation sites in chromatin utilized *E. coli* RNA polymerase. It was thus of interest to determine whether the sites utilized by the homologous hen oviduct RNA polymerase II corresponded to those used by the bacterial RNA polymerase, since serious doubts could be raised about the biological validity of using a procaryotic enzyme to transcribe chromatin.

Few meaningful studies on the initiation specificities of bacterial and eucaryotic RNA polymerases in transcribing chromatin are available. The most direct approach to study the fidelity of *in vitro* chromatin transcription has been the use of complementary DNA probes to search for specific RNA products. *E. coli* RNA polymerase is able to transcribe globin mRNA sequences from chromatin isolated from various hematopoetic tissues (Axel *et al.*, 1973; Gilmour and Paul, 1973; Steggles *et al.*, 1974; Barrett *et al.*, 1974). Chromatin from nonhematopoetic tissues or deproteinized DNA did not direct the transcription of detectable globin mRNA sequences. Likewise virus specific RNA has been identified in the bacterial RNA polymerase synthesized transcripts of chromatin from cells transformed by SV40 (Astrin, 1973; Shih *et al.*, 1973) and from cells infected with avian myeloblastosis virus (Jacquet *et al.*, 1974). Using homologous RNA polymerase II to transcribe chromatin, Steggles *et al.* (1974) found that globin mRNA sequences were present in a slightly higher proportion (about 3 times) than in transcripts synthesized by bacterial RNA polymerase. Jacquet *et al.* (1974) found, however, that both eucaryotic and bacterial RNA polymerases transcribed viral mRNA sequences to roughly the same extent. These studies emphasize that the tissue specific restriction of DNA sequences in chromatin is maintained during normal isolation procedures. Little information on the accuracy of transcription, however, is provided. The location of initiation sites for RNA synthesis utilized by the exogenously added RNA polymerases cannot be examined by this technique.

To experimentally determine whether different RNA polymerases utilized common or unique sites for initiation in chromatin, the ribonucleotide-rifampicin challenge technique for determining the number of available initiation sites was utilized. The results of a competition experiment between *E. coli* holoenzyme and core enzyme is shown in Figure 6. Enzyme saturation curves of fully stimulated chick oviduct chromatin were determined for holoenzyme alone, core enzyme alone, or an equimolar mixture of the two polymerases. From the saturation curves obtained for either enzyme alone, theoretical curves were calculated for both the case of complete competition between the two enzymes (curve D) and the case of no competition between the two polymerases (curve C). It can clearly be seen that the experimental points for the mixture of holo- and core enzymes followed closely the theoretical curve expected if both enzymes used common sites. Therefore, with or without sigma factor, *E. coli* RNA polymerase utilized, in general, the same set of initiation sites in chick oviduct chromatin.

The results of a similar competition experiment between *E. coli* holoenzyme and hen oviduct RNA polymerase II are shown in Figure 7. For this experiment, the inhibitor rifamycin AF/013 was used in place of rifampicin.

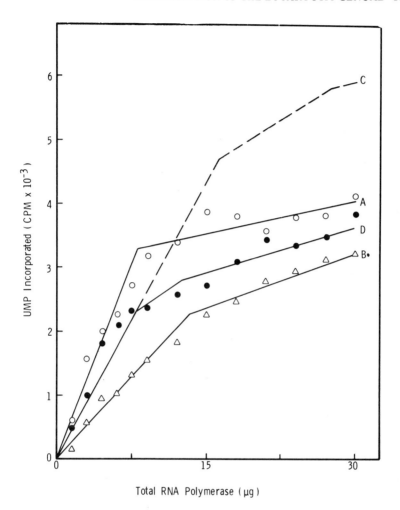

Fig. 6. *RNA initiation sites utilized by* E. coli *RNA polymerase holoenzyme and core enzyme on chick oviduct chromatin.* E. coli *holoenzyme (○, curve A), core enzyme (△ curve B), or a mixture of the two enzymes in equal concentrations (●) was incubated at 37°C with 5.0 µg of chick oviduct chromatin (14 day DES stimulated) in 0.2 ml of preincubation buffer for 40 minutes. RNA synthesis was initiated by the addition of 0.05 ml of ribonucleoside triphosphate mixture containing 0.2 mg/ml rifampicin and 4.0 mg/ml heparin and carried out for 1.5 minutes at 37°C. Theoretical curves were constructed from the experimental curves A and B assuming either no competition (curve C) or complete competition (curve D) between the two polymerases for RNA initiation sites.*

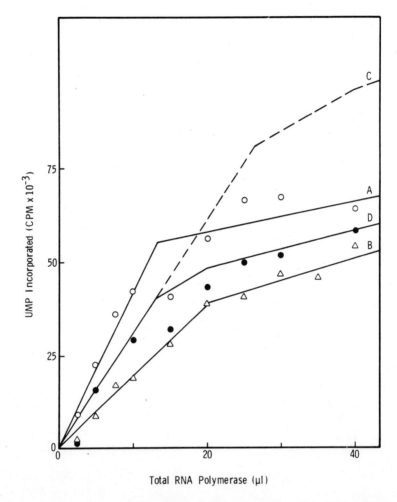

Fig. 7. *RNA initiation sites utilized by* E. coli *holoenzyme and hen oviduct RNA polymerase II on chick oviduct chromatin. Hen oviduct RNA polymerase II (○, curve A), * E. coli *holoenzyme (△, curve B), or a mixture of the two enzymes in equal volumes (●) was incubated at 37°C with 5.0 μg of chick oviduct chromatin (14 day DES stimulated) in 0.2 ml of preincubation buffer for 15 minutes. Stock concentrations used were 130 units/ml for hen RNA polymerase and 400 μg/ml for holoenzyme. After 15 minutes, RNA synthesis was initiated by the addition of 0.05 ml of a mixture containing 3.0 mM each of ATP, CTP, and GTP, 0.3 mM [³H]UTP (0.4 mC/ml) and 1 mg/ml rifamycin AF/013. RNA synthesis was carried out for 15 minutes at 37°C. Theoretical curves were constructed from experimental curves A and B assuming either no competition (curve C) or complete competition (curve D) of the two polymerases for RNA initiation sites. Hen oviduct RNA polymerase II was purified by a modified procedure of Kedinger* et al. *(1972a,b).*

Although not as well characterized as rifampicin, rifamycin AF/013 has been shown to inhibit eucaryotic RNA polymerase II prior to the formation of the highly stable binary complex (Meilhac et al., 1972; Tsai and Saunders, 1974). Theoretical curves were again constructed for the two extreme cases of complete competition between the two enzymes (curve D) and no competition between the two polymerases (curve C). The experimental points for the mixture of holoenzyme and hen oviduct RNA polymerase closely followed the curve predicted if both enzymes used common sites. Thus, regardless of the RNA polymerase used, RNA synthesis appeared to be initiated from the same region of the chromatin. This is consistent with the hypothesis that certain structural features of the chromatin may help to determine the proper sites for initiation. It should be pointed out, however, that at the present time we have no evidence to indicate that the initiation sites we have measured *in vitro* correspond to the actual promoter sites used for initiation *in vivo*.

Although the above experiments indicated that different RNA polymerases initiate from a set of common sites, they do not distinguish the efficiency with which these sites are utilized by different RNA polymerases. In the presence of rifampicin or rifamycin AF/013, each initiation site is limited to a single initiation event. To determine whether RNA polymerases transcribed from a given site with varying efficiencies, RNA synthesis was carried out in the absence of any initiation inhibitor. The RNA synthesized with chromatin isolated from an estrogen-stimulated chick oviduct was tested for the presence of sequences homologous to ovalbumin mRNA by hybridization to a complementary DNA probe. The amount of ovalbumin mRNA sequences present in the chromatin transcripts was determined by comparison of the initial rate of hybridization to the rate determined for purified ovalbumin mRNA. As shown in Figure 8, hen oviduct RNA polymerase II transcribed more ovalbumin mRNA sequences (0.04% of total RNA) than either *E. coli* holoenzyme or core enzyme. The relative ratios of ovalbumin mRNA sequences transcribed by oviduct RNA polymerase, core enzyme, and holoenzyme were 4:1.5:1, respectively. Control experiments in which chromatin was incubated without added exogenous RNA polymerase indicated that the level of contaminating ovalbumin mRNA was insignificant. Therefore, the hybridization observed was due to the transcription of newly synthesized sequences of ovalbumin mRNA. Hen oviduct RNA polymerase II seems to be more efficient at transcribing RNA from sequences homologous to ovalbumin mRNA than either core or holo polymerase from *E. coli*. Furthermore, RNA transcripts of chromatin isolated from an unstimulated chick oviduct or of deproteinized chick DNA did not contain detectable ovalbumin mRNA sequences (Harris et al., 1975b). Thus, chromatin isolated from chick oviduct maintains the differentiation stage-specific pattern of restriction observed *in vivo*.

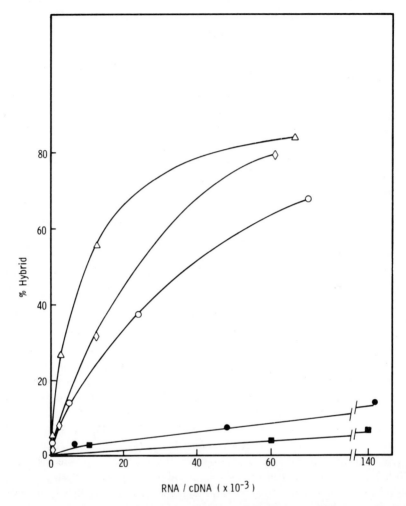

Fig. 8. *Presence of sequences homologous to ovalbumin mRNA in chick oviduct chromatin transcripts. RNA was synthesized in 2.5 ml of solution containing 50 mM Tris-HCl, pH 7.9, 1 mM $MnCl_2$, 2 mM 2-mercaptoethanol, 50 mM $(NH_4)_2SO_4$, 4 µg/ml α-amanitin, 300 µg of chick oviduct chromatin (14 day DES stimulated) and either 800 µg of E. coli RNA polymerase holoenzyme (○), 1600 µg of core enzyme (◊), or 400 units of hen oviduct RNA polymerase II (△). Controls were run in which 300 µg of chick DNA was used in place of chromatin with holoenzyme (●) and in which no exogenous RNA polymerase was added (■). Reactions were incubated for 2 hours at 37° and then centrifuged at 12,000 × g for 20 minutes to remove chromatin. RNA was extracted essentially by the procedure of Harris et al. (1975b). RNA was hybridized to complementary DNA transcribed from ovalbumin mRNA by the procedure of Harris et al. (1975a). Each hybridization reaction contained 1.5 ng of cDNA and was carried out for 96 hours in 0.05 ml of 0.01 M Tris-HCl, pH 7.0, 0.6 M NaCl at 68°C. The amount of cDNA resistant to S_1 nuclease was determined as previously described (Harris et al., 1975a).*

V. PROPERTIES OF *IN VITRO* TRANSCRIPTION OF CHROMATIN

A. Determination of the Rate Limiting Step of RS Complex Formation on Chromatin

As discussed earlier, the sequence leading to the formation of the RS complex involves the following steps: (1) dissociation of enzyme from nonspecific binding sites; (2) binding of the enzyme at an initiation site to form the I complex; (3) conversion of the I complex to the highly stable RS complex through a local denaturation of DNA. Any of these steps can be rate-limiting in the formation of RS complex. For chromatin, the $t_{1/2}$ of RS complex formation was seven times longer than that for DNA. To better understand this difference between chromatin and DNA, we have attempted to determine which step in the formation of RS complexes is rate limiting to the overall process.

The effect of the temperature of preincubation on RS complex formation can be informative in attempting to determine the rate-limiting step. The transition from I to RS complex (step 3) involves local opening of the DNA strands and is thus strongly dependent on temperature. Although we have no direct data on the temperature dependencies of steps (1) and (2), we assume from the nature of the interactions involved that these steps would not be highly dependent on temperature. Mangel and Chamberlain (1974) have demonstrated that the formation of I complex between *E. coli* RNA polymerase and T_7 DNA is not markedly influenced by temperature. Therefore, if step (3) is rate limiting, decreasing the temperature of preincubation should increase the $t_{1/2}$ of RS complex formation. On chick DNA this was the case (Figure 9A). The $t_{1/2}$ of RS complex formation increased from 1.3 to 8.3 minutes as the temperature of preincubation decreased from 37°C to 0°C. Concomitantly, the maximum level of RNA synthesis observed (RS max) decreased in a reciprocal fashion as the temperature of preincubation was lowered. The strong temperature dependency of RS max is consistent with the requirement for opening of the DNA strands for RS complex formation. The temperature dependency of $t_{1/2}$ suggests that at lower temperatures the conversion from I to RS complex is the rate-limiting step in formation of RS complexes on DNA. At 37°, the rate-limiting step on DNA has been shown to be step (1), the dissociation of enzyme from nonspecific binding sites (Hirose *et al.*, 1975).

When chick oviduct chromatin was used as a template, the effects of temperature on RS complex formation were dramatically different (Figure 9B). The dependency of RS max on the temperature of preincubation was relatively poor compared to DNA (note scale change). Thus, the local opening of the DNA strands during the conversion of I to RS complex is not as highly dependent on temperature as that seen for DNA. Furthermore, the $t_{1/2}$ of RS

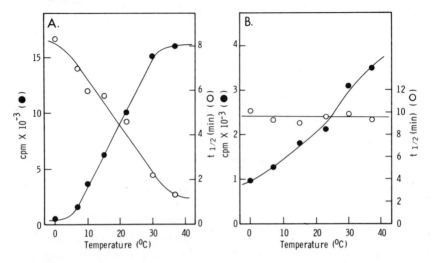

Fig. 9. *Temperature dependency of RS complex formation. (A)* E. coli *RNA polymerase (1.4 µg) was incubated with chick DNA (0.75 µg) in 0.1 ml of preincubation buffer at the indicated temperature. At various time periods, RNA synthesis was initiated by the addition of 0.025 ml of ribonucleoside triphosphate mixture and carried out for 15 minutes at 37°C. The maximum level of [³H] UTP incorporation (●) and $t_{1/2}$ (○) were determined as shown in Figure 1. (B)* E. coli *RNA polymerase (7.0 µg) was incubated with 5.0 µg of chick oviduct chromatin (14 day DES stimulated) at the indicated temperature. After various time periods, RNA synthesis was initiated by the addition of 0.05 ml of ribonucleoside triphosphate mixture. Other conditions were as described above.*

complex formation on chromatin was completely independent of preincubation temperature. Therefore, the conversion from I to RS complex (step 3) is not the rate-limiting step in RS complex formation on chromatin throughout the temperature range tested.

The formation of I complex (step 2) can also be eliminated as a possible rate-limiting step in RS complex formation on chromatin. If step 2 were rate-limiting, the formation of RS complexes should occur via a second order process. As indicated in Figure 1, however, RS complex formation occurs via a first order process. Furthermore, varying the concentration of RNA polymerase and chromatin (as DNA) at a fixed ratio had no effect on the $t_{1/2}$ of RS complex formation. Therefore, neither step 2 nor step 3 is rate limiting in the formation of RS complex on chromatin.

Consequently, the rate-limiting step in the formation of RS complexes on chromatin appears to be the dissociation of enzyme from nonspecific interactions (step 1). These interactions could be with the DNA itself or with chromosomal proteins. The nature of these nonspecific interactions was more closely examined by testing the effect of various agents on the $t_{1/2}$ of RS complex formation. Changing the ammonium sulfate concentration in the preincubation mixture from 62.5 mM to 0 had little effect on the $t_{1/2}$ of RS complex formation on chromatin (Table III). This would not be expected if ionic interactions between RNA polymerase and chromatin were determining the rate-limiting step. Thus, the interaction between enzyme and chromatin may be hydrophobic in nature. To test this possibility, the effect of 1-anilino-8-naphthalene sulfonate (ANS) on the $t_{1/2}$ of RS complex formation was tested (Table III). ANS is known to bind to hydrophobic regions of proteins and thus may interfere with hydrophobic interactions between proteins. The $t_{1/2}$ was measured at concentrations of ANS which had no effect on the total RNA polymerase activity. In the presence of 0.5 mM ANS, the $t_{1/2}$ of RS complex formation on chromatin decreased by 3.9 minutes. The same concentration of ANS did not alter the $t_{1/2}$ of RS complex formation on DNA. Thus, it is possible that hydrophobic interactions with chromosomal proteins may play a major role in the rate at which RNA polymerase can locate initiation sites in the chromatin. These nonspecific interactions cause the rate of formation of RS complexes on chromatin to be significantly slower than RS complex formation on deproteinized DNA.

B. Role of Histone and Nonhistone Proteins in the Initiation of Transcription

The low temperature dependency of RS complex formation on chromatin compared to DNA must be due, directly or indirectly, to the presence of chromosomal proteins. It was of interest to determine whether any specific subfraction of chromosomal proteins was responsible for this characteristic of chromatin transcription. In this regard, we first separated chromosomal proteins into histone and nonhistone fractions using Bio Rex 70 chromatography (Parker et al., 1975). These fractions were then reconstituted with

TABLE III.
Effects of Varying Conditions of Preincubation on RS Complex Formation[a]

Conditions	Chick Oviduct Chromatin	Chick DNA (Native)	Chick DNA (Denatured)
$t_{1/2}$ (62.5 mM AmSO$_4$; 37°)	9.4	1.3	1.5
$t_{1/2}$ (0 mM AmSO$_4$; 37°)	10.9	2.7	---
RS max (+AmSO$_4$)/RS max (−AmSO$_4$)	1.1	1.1	---
$t_{1/2}$ (62.5 mM AmSO$_4$; 37°; 0.5 mM ANS)	5.5	1.3	---
RS max (+ANS)/RS max (−ANS)	1.0	1.0	---
$t_{1/2}$ (62.5 mM AmSO$_4$; 0°C)	10.1	8.3	1.9
RS max (37°)/RS max (0°)	3.5	36	2.2

[a]RS max and $t_{1/2}$ were determined as described in Figure 1 with the noted changes in the preincubation buffer or temperature.

chick DNA by the method of Stein *et al.* (1975) and the effect of temperature of preincubation on the level of RNA synthesis was tested. The transition temperature is defined as the temperature of preincubation at which the level of RNA synthesis is 50% of the difference between the levels of RNA synthesis after preincubation at 37° and 0°. For native DNA, a transition temperature should correspond to the temperature at which 50% of the initiation sites are in the opened form (RS complex) and 50% are in the closed form (I complex).

Native chick DNA had a transition temperature of 24° under the conditions tested (Figure 10). When chick DNA was reconstituted with

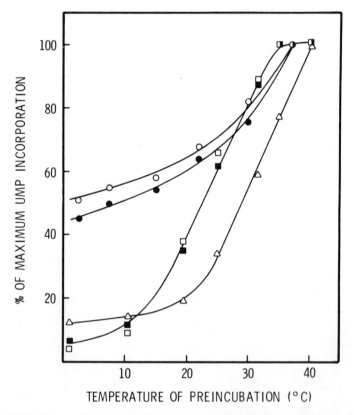

Fig. 10. *Effect of histones and nonhistones on transition temperature of RS complex formation.* E. coli *RNA polymerase (10 μg) was incubated with 5 μg of template in 0.2 ml of preincubation buffer at the temperature indicated. After 40 minutes, RNA synthesis was initiated by the addition of 0.05 ml of ribonucleoside triphosphate mixture and carried out for 15 minutes at 37°C. The maximum level of [³H] UTP incorporation was normalized to 100% for each individual template. Templates used and maximum levels of RNA synthesized were:* □, *DNA (184,000 CPM);* ■, *nonhistone-DNA complex (129,000 CPM);* △, *histone-DNA complex (7,900 CPM);* ●, *nonhistone-histone-DNA complex (total reconstitute, 7,600 CPM); and* ○, *native chromatin (6,800 CPM).*

histones at a mass ratio of 1:1, the transition temperature rose to about 30°. This estimate is a lower limit of the true transition temperature as the level of RNA synthesis from the histone-DNA complex had not reached a maximum at 37°. The level of RNA synthesis after preincubation at 37° for the histone-DNA complex was only 4 to 5% of that for deproteinized DNA. Histones, thus, appear to be general repressors of RNA synthesis which act to stabilize the DNA duplex structure against formation of the RS complex.

Chick DNA was reconstituted with the total nonhistone fraction at a mass ratio of 1:2, respectively. These nonhistone-DNA complexes had a transition temperature which was not significantly different from that of DNA alone (Figure 10). The level of RNA synthesis after preincubation at 37° was also similar for the nonhistone-DNA complex and deproteinized DNA. Neither the total nonhistone fraction nor the histone fraction alone, therefore, were able to alter the characteristics of DNA transcription to cause a temperature profile which resembled that for native chromatin.

The effect of the temperature of preincubation on the total chromatin reconstitute between histones, nonhistones, and DNA was next examined. The total reconstituted chromatin resembled very closely the temperature profile of native chromatin (Figure 10). The level of RNA synthesis after preincubation at 37° was approximately 3% of that observed for DNA and thus was similar to the level seen with histone-DNA complex. The temperature profile, however, is quite different from the histone-DNA complex. Even after preincubation at 0°, there is a fairly high level of RNA synthesis — about 40% of that observed after preincubation at 37°. Furthermore, the temperature profile of the remaining 60% of the activity was not as highly cooperative as the profiles seen for DNA, nonhistone-DNA, or histone-DNA complex. It is not possible to assign a transition temperature to either native or reconstituted chromatin due to the noncooperative nature of the curves. Nevertheless, it is clear that the low temperature dependency of RS complex formation on chromatin requires the presence of both histone and nonhistone proteins and cannot be attributed to any specific subfraction of chromosomal proteins.

C. Implications on the Structure of Initiation Sites in Chromatin

The effects of temperature on RS complex formation on chromatin appear very similar in character to that found for denatured DNA. As opposed to native, double-stranded DNA, the $t_{1/2}$ of RS complex formation and the level of RS max on denatured DNA showed little temperature dependency (Table III). This is most likely due to the fact that, on single stranded DNA, no opening of the DNA strands is necessary for the formation of a stable binary complex capable of rapidly initiating RNA synthesis. Thus, the process of RS complex formation on denatured DNA is not greatly affected by the temperature of preincubation.

The similarity of the temperature dependencies of RS max and $t_{1/2}$ for chromatin and denatured DNA suggested that initiation sites on chromatin could be regions in which the DNA is present in a single stranded structure. Crick (1971) has proposed that in higher organisms RNA initiation occurs at single stranded regions of DNA. To test this possibility, the effect of single strand specific nuclease and antibody to single stranded DNA on chromatin transcription were examined. If significant regions of single stranded DNA structure exist at initiation sites in chromatin, these proteins should be highly efficient at inhibiting *in vitro* transcription from chick oviduct chromatin.

The effects of antibody against both single stranded DNA (A) and double stranded DNA (B) are shown in Figure 11. Antibody against single stranded DNA was highly effective at inhibiting transcription from denatured chick DNA, presumably by binding to the DNA and blocking RNA polymerase. Transcription from native chick DNA and chromatin, however, were not greatly affected by the single strand specific antibody except at extremely high titers. Antibody to double stranded DNA was efficient at completely

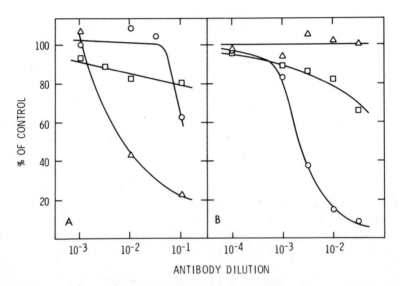

Fig. 11. *Effect of antibody against DNA on rifampicin resistant RNA synthesis. Antibody against either single stranded DNA (A) or double stranded DNA (B) was incubated at 25°C with template in 2 mM Tris-HCl, pH 7.9, 1 mM EDTA at the indicated titers. After 1 hour, 7.0 µg of E. coli RNA polymerase was added together with preincubation buffer to a volume of 0.2 ml. After 15 minutes of further incubation at 37°C, RNA synthesis was initiated by the addition of 0.05 ml of ribonucleoside triphosphate mixture and carried out for 15 minutes. Antibody against single stranded DNA was the gift of Dr. Lawrence Levine and against double stranded DNA of Dr. B. D. Stoller. Templates used and levels of [^{3}H] UTP incorporation in the absence of antibody were:* ○, *1.5 µg of native chick DNA (79,000 CPM);* △, *1.5 µg of denatured chick DNA (9,900 CPM); and* □, *5 µg of 14 day DES stimulated chromatin (5,600 CPM).*

inhibiting transcription from native chick DNA at titers which had no effect on transcription of denatured DNA. The transcription of chromatin was only slightly inhibited by the antibody to double stranded DNA. Thus, neither antibody to single or double stranded DNA was very efficient at inhibiting the transcription of chromatin by RNA polymerase. The probable explanation for this observation is that the initiation of transcription does not occur at regions of 'naked DNA'. Therefore, the antibody molecules would not be efficiently able to penetrate the chromosomal proteins to interact with the chromosomal DNA to inhibit transcription.

The effect of the Mung Bean single strand specific nuclease was also examined (Figure 12). This protein is much smaller than antibody and thus might be better able to penetrate the chromatin structure. Pretreatment of denatured chick DNA with as little as 0.02 units of Mung Bean nuclease was sufficient to cause almost complete inactivation of this template. Similar treatment of native chick DNA or chromatin showed little inactivation. Even when chromatin was treated with as much as 0.2 units of nuclease, ten times higher than that necessary to completely inactivate denatured DNA, no inhibition of transcription was detected. These studies indicate that regions of single stranded DNA which are large enough to interact with nuclease and are actively transcribed by *E. coli* RNA polymerase do not exist in chromatin. More likely, initiation sites in chromatin are associated with chromosomal proteins which block the binding of either antibody or nuclease molecules. These chromosomal proteins must somehow interact with the DNA to facilitate the opening of the DNA strands during formation of the RS complex between RNA polymerase and the promoter site.

VI. SUMMARY

The chick oviduct can be induced by the administration of steroid hormones to undergo dramatic cellular changes which appear to be mediated at the level of transcription. Thus, the chick oviduct is an excellent system for studying the regulation of gene expression in a eucaryotic organism. Our approach has been to utilize chromatin isolated from chick oviduct as a template for *in vitro* transcriptional studies. To understand the mechanism of estrogen-mediated alteration of gene expression, we first felt it necessary to more fully characterize the interaction between chromatin and purified RNA polymerase.

The formation of binary complexes capable of rapidly initiating RNA synthesis (RS complexes) between *E. coli* RNA polymerase and chromatin differs in several respects from the formation of such complexes on deproteinized DNA. The half time of RS complex formation on chromatin was approximately seven times slower than that for DNA. The rate limiting step in

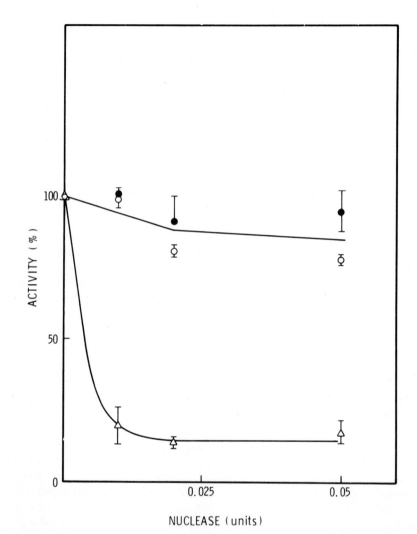

Fig. 12. *Effect of Mung Bean nuclease on rifampicin resistant RNA synthesis. Various amounts of Mung Bean nuclease were incubated with template at 37°C for 30 minutes in 0.02 M sodium acetate, pH 5.5 After neutralization with NaOH, 7.0 µg of E. coli RNA polymerase together with preincubation buffer were added to a volume of 0.2 ml. After 15 minutes at 37°C, RNA synthesis was initiated by the addition of 0.05 ml of ribonucleoside triphosphate mixture and carried out for 15 minutes. Mung Bean nuclease was the gift of Dr. M. Laskowski, Sr. Templates used and levels of [³H]UTP incorporation in the absence of nuclease were:* ○, *1.5 µg of native chick DNA (72,000 CPM);* △, *1.5 µg of denatured chick DNA (9,700 CPM); and* ●, *5 µg of 14 day DES stimulated chromatin (7,400 CPM).*

the formation of RS complexes on chromatin appears to be the rate of dissociation of enzyme from nonspecific binding sites in the chromatin. Studies with ANS suggested that these interactions may be largely hydrophobic in nature. Thus, the presence of chromosomal proteins tends to retard the rate by which RNA polymerase locates an initiation site on chromatin. Such random, reversible interactions, however, could conceivably serve a biological function *in vivo* to trap the RNA polymerase molecules in the proximity of chromosomal DNA.

As opposed to the formation of RS complexes, the rate of RNA chain initiation from preformed RS complexes is a fairly rapid event. Furthermore, the rate constants of RNA chain initiation for chromatin and DNA are not remarkably different. Thus, once the RS complex is formed, the actual event of RNA chain initiation may be quite similar for the two templates.

The formation of RS complexes between RNA polymerase and chromatin was much less dependent on either the temperature of preincubation or the presence of sigma factor than that for deproteinized DNA. Since these two agents are thought to be involved in the opening of the DNA strands, initiation sites in chromatin may be present in a state in which conversion of I to RS complex is greatly facilitated. One possible explanation is that the initiation regions of chromatin are stretches of single stranded DNA. The presence of such regions in chromatin accessible to RNA polymerase seems unlikely, however, in view of the lack of inhibition of chromatin transcription by antibody to single stranded DNA or single strand specific nuclease. A more probable explanation is that chromosomal proteins present at initiation sites lower the energy required for opening of the DNA strands in formation of the RS complex. Such proteins would be expected to be major determinants of the location of initiation sites for RNA polymerase. Competition experiments with *E. coli* holoenzyme, core enzyme, and hen oviduct RNA polymerase II indicate that these polymerases all utilize, to a large extent, the same set of initiation sites on chromatin. Since these enzymes have different specificities on native DNA (Bautz and Bautz, 1970; Meilhac and Chambon, 1973), it seems likely that chromosomal proteins are influencing the selection of initiation sites on chromatin.

The total number of initiation sites available in chromatin for RNA polymerase increased after primary or secondary estrogenic stimulation of the chick oviduct. Simultaneous measurements of the rate of RNA chain elongation and RNA chain length were fairly constant throughout the period of estrogen administration. Thus, increases in the "template capacity" of chromatin measured previously reflected an increase in the available DNA sequences for RNA polymerase transcription. During primary stimulation, a maximum level of initiation sites, seven-fold higher than the unstimulated chick, was reached at about four days of stimulation. This level then decreased to a plateau level about two-times higher than the unstimulated animal by about ten days of stimulation. The high level of initiation sites at four days corresponds with the period when maximal differentiation of cell types and growth are occurring. Thus, it appears that cells in the actual process of differentiation may be transcribing a much larger number of genes than either the undifferentiated or highly differentiated cell types. Following completion of the differentiation process, the need for this increased expression of genetic information appears to subside, since the number of initiation sites for RNA synthesis decreases.

During the secondary stimulation of chick oviduct with estrogen, the level of nuclear receptors for estrogen increased immediately prior to an increase in the number of initiation sites, which in turn preceded an increase in the level of intracellular ovalbumin mRNA. This temporal relationship suggested that a possible direct interaction of hormone-receptor complex with chromatin may be involved in opening new DNA sequences for transcription. To test this possibility, we are currently examining the effects of purified estrogen receptor on *in vitro* oviduct chromatin transcription. By using our *in vitro* assay for the number of chromatin initiation sites and the complementary DNA probe for ovalbumin mRNA, we are able to monitor for expected changes in the chromatin of unstimulated chicks. Other studies are aimed at better defining the roles of various nonhistone proteins in transcription. Reconstitution of DNA and histones with nonhistones from various developmental stages will provide an assessment of the role of nonhistones in determining the specificity of initiation. We are also currently in the process of purifying the ovalbumin gene from total cellular DNA. With the purified gene for ovalbumin, we would be able to study the molecular mechanisms of control of a specific gene *in vitro* and attempt to reconstitute the *in vivo* regulation from purified components of the transcription apparatus.

ACKNOWLEDGEMENTS

This work was supported by NIH grants HD-8188, HD-7857 and HD-7495 (Center grant for Population Research and Reproductive Biology) and a grant from the Ford Foundation.

REFERENCES

Anderson, J., Clark, J. H., and Peck, Jr., E. J. (1972). *Biochem. J.* **126**, 561–567.
Astrin, S. M. (1973). *Proc. Nat. Acad. Sci. U.S.A.* **70**, 2304–2308.
Axel, R., Cedar, H., and Felsenfeld G. (1973). *Proc. Nat. Acad. Sci. U.S.A.* **70**, 2029–2032.
Barrett, T., Maryanka, D., Hamlyn, P. H., and Gould, H. (1974). *Proc. Nat. Acad. Sci. U.S.A.* **71**, 5057–5061.
Bautz, E. K. F., and Bautz, F. (1970). *Nature* **226**, 1219–1222.
Burgess, R. R., and Travers, A. A. (1971). *In* "Methods in Enzymology" (L. Grossman and K. Moldave, eds.), Vol. XXI(D), pp. 500–506. Academic Press, New York.

Chamberlin, M. J. (1974). *Ann. Rev. Biochem.* **43**, 721–775.
Chamberlin, M. J., and Ring, J. (1972). *J. Mol. Biol.* **70**, 221–237.
Chan, L., Means, A. R., and O'Malley, B. W. (1973). *Proc. Nat. Acad. Sci. U.S.A.* **70**, 1870–1874.
Comstock, J. P., Rosenfeld, G. C., O'Malley, B. W., and Means, A. R. (1972). *Proc. Nat. Acad. Sci. U.S.A.* **69**, 2377–2380.
Cox, R., Haines, M., and Carey, N. (1973). *Eur. J. Biochem.* **32**, 513–524.
Cox, R. F., Haines, M. E., and Emtage, J. S. (1974). *Eur. J. Biochem.* **49**, 225–236.
Crick, F. (1971). *Nature* **234**, 25–27.
Gilmour, R. S., and Paul, J. (1973). *Proc. Nat. Acad. Sci. U.S.A.* **70**, 3440–3442.
Harris, S. E., Means, A. R., Mitchell, W. M., and O'Malley, B. W. (1973). *Proc. Nat. Acad. Sci. U.S.A.* **70**, 3776–3780.
Harris, S. E., Rosen, J. M., Means, A. R., and O'Malley, B. W. (1975a). *Biochemistry* **14**, 2072–2081.
Harris, S. E., Schwartz, R. J., Tsai, M.-J., Roy, A. K., and O'Malley, B. W. (1975b). *J. Biol. Chem.* In press.
Hinkle, D. C., Mangel, W. F., and Chamberlin, M. J. (1972). *J. Mol. Biol.* **70**, 209–220.
Hirose, M., Tsai, M.-J., and O'Malley, B. W. (1975). *J. Biol. Chem.* In press.
Jacquet, M., Groner, V., Monroy, G., and Hurwitz, J. (1974). *Proc. Nat. Acad. Sci. U.S.A.* **71**, 3045–3049.
Kedinger, C., and Chambon, P. (1972b). *Eur. J. Biochem.* **28**, 283–290.
Kedinger, C., Gissinger, F., Gniazdowski, M., Mandel, J., and Chambon, P. (1972a). *Eur. J. Biochem.* **28**, 269–276.
Lewin, B. (1974). *In* "Gene Expression, Volume I, Bacterial Genomes" John Wiley and Sons, London.
Mangel, W. F., and Chamberlin, M. J. (1974). *J. Biol. Chem.* **249**, 2995–3001.
Means, A. R., Comstock, J. P., Rosenfeld, G. C., and O'Malley, B. W. (1972). *Proc. Nat. Acad. Sci. U.S.A.* **69**, 1146–1150.
Meilhac, M., and Chambon, P. (1973). *Eur. J. Biochem.* **35**, 454–463.
Meilhac, M., Tysper, Z., and Chambon, P. (1972). *Eur. J. Biochem.* **28**, 291–300.
O'Malley, B. W., and McGuire, W. L. (1968). *Proc. Nat. Acad. Sci. U.S.A.* **60**, 1527–1534.
O'Malley, B. W., McGuire, W. L., Kohler, P. O., and Korenman, S. G. (1969). *Rec. Prog. Horm. Res.* **25**, 105–160.
O'Malley, B. W., and Means, A. R. (1974). *Science* **183**, 610–620.
Palmiter, R. D. (1973). *J. Biol. Chem.* **248**, 8260–8270.
Palmiter, R. D., Christensen, A. K., and Schimke, R. T. (1970). *J. Biol. Chem.* **245**, 833–845.
Parker, M. G., Sheehan, D. M., and O'Malley, B. W. (1975). Submitted.
Rhoads, R. E., McKnight, G. S., and Schimke, R. T. (1973). *J. Biol. Chem.* **248**, 2031–2039.
Schwartz, R. J., Tsai, M.-J., Tsai, S. Y., and O'Malley, B. W. (1975). *J. Biol. Chem.* **250**, 5175–5182.
Shih, T. Y., Khoury, G., and Martin, M. A. (1973). *Proc. Nat. Acad. Sci. U.S.A.* **70**, 3506–3510.
Sippel, A., and Hartman, G. (1968). *Biochim. Biophys. Acta* **157**, 218–219.
Spelsberg, T. C., Mitchell, W. M., Chytil, F., Wilson, E. M., and O'Malley, B. W. (1973). *Biochim. Biophys. Acta* **312**, 765–768.
Steggles, A. W., Wilson, G. N., Kantos, J. A., Picciano, D. J., Falvey, A. K., and Anderson, W. F. (1974). *Proc. Nat. Acad. Sci. U.S.A.* **71**, 1219–1223.
Stein, G. S., Mans, R. J., Gabbay, E. J., Stein, J. L., Davis, J., and Adawadkar, P. D. (1975). *Biochemistry* **14**, 1859–1866.

Tsai, M.-J., and Saunders, G. F. (1973). *Proc. Nat. Acad. Sci. U.S.A.* **70,** 2072–2076.
Tsai, M.-J., Schwartz, R. J., Tsai, S. Y., and O'Malley, B. W. (1975). *J. Biol. Chem.* **250,** 5165–5174.
Tsai, S. Y., Tsai, M.-J., Schwartz, R. J., Kalimi, M., Clark, J. H., and O'Malley, B. W. (1975). *Proc. Nat. Acad. Sci. U.S.A.* In press.
Umezawa, H., Mizuno, S., Yamazaki, H., and Nitta, K. (1968). *J. Antibiotics* **21,** 234–235.

ESTROGEN-RECEPTOR MOVEMENT TO THE NUCLEUS: DISCUSSION OF A CYTOPLASMIC-EXCLUSION HYPOTHESIS

Frank Gannon[1]
*Department of Biochemistry,
University of Wisconsin,
Madison, Wisconsin 53706*

Benita Katzenellenbogen
*Departments of Physiology & Biophysics,
University of Illinois
Urbana, Illinois 61801*

George Stancel
*Program in Pharmacology,
University of Texas Medical School at Houston,
Texas Medical Center,
Houston, Texas 77025*

Jack Gorski
*Department of Biochemistry,
University of Wisconsin,
Madison, Wisconsin 53706*

I. INTRODUCTION

Fundamental to the action of estrogens in target tissues is the binding of the hormone to a receptor-protein and the subsequent movement of the hormone-receptor complex from the cytoplasm into the nucleus (Gorski *et al.*, 1968; Shyamala and Gorski, 1969; Jensen *et al.*, 1968). This is perhaps the only aspect of steroid hormone action which is universally accepted and it appears to most to be a prerequisite to the alteration, by the hormone, of the genetic expression of the cell.

Molecular changes in the receptor which appear to be necessary for this translocation have been the subject of many studies (Stancel *et al.*, 1973a,b; Notides and Nielsen, 1974; Jensen *et al.*, 1971; Gschwendt and Hamilton, 1972), but only limited information is available concerning the actual intracellular movement of the receptor. In fact, this very important characteristic of the receptor has been consistently subordinate to speculation on the ultimate question; how does the estrogen-receptor complex alter the genetic expression of the cell. It has been assumed that the answer to that question

---
[1] Present address: Faculté de Medecine, Strasbourg, France

would automatically provide the reason for the translocation of the receptor. The rationale for this assumption becomes obvious if one accepts the following logic: Data from other systems indicate that a change in the genetic response of a cell results from variations in the manner in which proteins are deployed on the chromatin. The direct interaction of estrogen-receptor with some locus on the chromatin could cause such a perturbation. The specificity of the hormone response should then reflect a specific receptor-chromatin interaction with the chromatin segment involved in this scenario being called the "acceptor". These acceptor sites are generally believed, by analogy with other systems, to be few in number and to have a very high affinity for the receptor-hormone complex. In this model, the translocation of the receptors into the nucleus simply represents a movement in response to the high affinity of the estrogen-receptor complex for the acceptor sites.

It should be clear, however, that whereas the idea of such acceptors is a satisfying concept, it has been deduced and extrapolated from data which are frought with difficulties of interpretation. Evidence supporting the existence of acceptors in the nucleus might be summarized as: (a) the preponderance of the estrogen-receptor complex in the nucleus after incubation of target tissues at temperatures greater than 25°C (Jensen et al., 1968; Noteboom and Gorski, 1963); (b) the fact that KCl concentrations of greater than 0.3 M are required to remove the majority of the estrogen-receptor complexes from the nucleus (Jensen et al., 1967; Jungblut et al., 1967; Puca and Bresciani, 1968); (c) the isolation, from the nuclei of tissues which respond to the presence of estrogens, of proteins which bind the hormone-receptor complex with particular avidity and which therefore could stake the claim of being the "acceptor" (Spelsberg et al., 1971; Puca et al., 1974). In this paper we will examine possible explanations for the translocation of the estrogen-receptor to the nucleus (a, above). In doing so we present some new evidence concerning translocation and, building upon a hypothesis of Horowitz and Moore (1974), propose an explanation for translocation which is independent of the presence of acceptors.

II. METHODS

Immature female rats (21-24 days old) from Holtzman were used in all studies. To determine the distribution of the estrogen-receptor complex in the rat pituitary, ten anterior pituitaries were pooled for each sample. Duplicate samples were incubated at 37° for 1 hr in 2 ml of Eagle's tissue culture medium for HeLa cells (Difco) with either tritiated estradiol at concentrations from 1.75×10^{-10}M to 1.1×10^{-8}M or that concentration of tritiated estradiol plus one hundred fold excess of unlabeled estradiol. Following incubation, whole pituitaries were washed 5 times with 0.9% NaCl. Each sample was then suspended in 1.0 ml of Tris-HCl buffer pH 7.4 and homogenized by 25 handstrokes using a loose-fitting (A) Dounce homogenizer. The homogenate was passed through 4 layers of cheesecloth and the homogenizer and cheesecloth then washed with 2 ml of Tris-HCl buffer. The homogenate was then centrifuged at 800 g for 10 min at 0°C. The supernatant was retained for assay by Hydroxylapatite (Williams and Gorski, 1974) and the pellet for each sample resuspended in 1 ml of Tris-HCl buffer − 0.25 M sucrose pH 7.4. To ensure dispersal of the pellet, the samples were passed 10 times through a stainless steel #18 needle and 10 times through a #20 needle. The 1 ml samples were then layered onto a multi-step sucrose gradient which consisted of Tris-HCl buffer − 2.2 M sucrose, pH 7.4 (1 ml) Tris-HCl buffer − 1.23 M sucrose pH 7.4 (1 ml) and Tris-HCl buffer − 0.82 M sucrose, pH 7.4 (1 ml) in polyallomer tubes. These tubes were centrifuged at 75,000 g for 30 min (SW 56 swinging bucket, rotor, Beckman). The resultant pellet was resuspended as before in Tris-HCl buffer − 0.25 M sucrose pH 7.4, washed 3 times with this buffer and the bound estradiol determined by extraction with 2.0 ml of 100% ethanol. Nuclei prepared in this way appeared to be intact and only slightly contaminated by cell debris when examined under the light microscope. A measure of their "cleanliness" is the fact that the protein to DNA ratio was 0.98. Nuclear recovery, based on DNA analyses (Burton, 1956) was approximately 50%.

In the gradient studies to be described, uteri after incubation were homogenized directly in Tris-HCl (0.04 M), EDTA (.01 M) and KCl (0.4 M), pH 7.4 as described by Giannopoulos and Gorski (1971). The resultant homogenate was centrifuged at 50,000 x RPM (Spinco no. 50 rotor) for 45 minutes. An aliquot of the supernatant fraction was then layered on a 10-20% sucrose gradient containing 0.4 M KCl. Samples plus ^{14}C labeled ovalbumin as an internal marker were centrifuged at 50,000 x RPM (SW 56 rotor) for 12 hrs and 5 drop samples collected.

Studies on the *in vitro* induction of IP synthesis were carried out as described by Katzenellenbogen and Gorski (1972). Uteri were incubated as for the binding studies described above except that unlabeled estrogen was used. After 1 hr incubation, 30 µg/ml of actinomycin D was added to the medium plus either [^{14}C] leucine (controls) or [^{3}H] leucine (estrogen-treated) and the incubation continued for 2 hr. Cytosols were prepared, electrophoreses on polyacrylamide gels and the ratio of the isotopes in the IP band calculated as described previously (Katzenellenbogen and Gorski, 1972).

III. RESULTS AND DISCUSSION

Previous studies from our laboratory monitored the localization of the estrogen receptor in the rat uterus by following the distribution of radioactivity on fractionation of the uterus (Williams and Gorski, 1972). These results showed that at all estradiol concentrations the estrogen-receptor complex behaved as if in response to a simple equilibrium which, under *in vivo* conditions, had no saturable component. That this equilibrium situation is not peculiar to the uterus is demonstrated in Figure 1, which shows that incubation of immature rat pituitaries with varying concentrations of estradiol-17β results in a constant ratio of nuclear to cytoplasmic receptor at all levels of filled binding sites. In addition to using a different tissue, the present study also employed nuclei which were relatively uncontaminated, as opposed to the crude 800 g pellet for rat uteri. This may explain numerical differences in the nuclear to cytoplasmic ratio in the two tissues (5:1 uterus and 2:1 pituitary) though other possibilities also exist.

In both this study and the initial report by Williams and Gorski (1972), the distribution of the receptor has been examined using a direct assay of the receptor after tissue homogenization. Nuclear and cytoplasmic forms of the receptor can also be distinguished on the basis of their different sedimentation coefficients on sucrose density gradients; in 0.4 M salt the cytoplasmic receptor has a sedimentation coefficient of approximately 4S and the nuclear receptor sediments as a 5.6S entity (Chamness and McGuire, 1972). When tissue is directly homogenized in 0.4 M KCl, without prior subcellular fractionation, the nuclear and cytoplasmic forms of the receptor are extracted as 4S and 5.6S forms respectively into the medium and subsequent centrifuga-

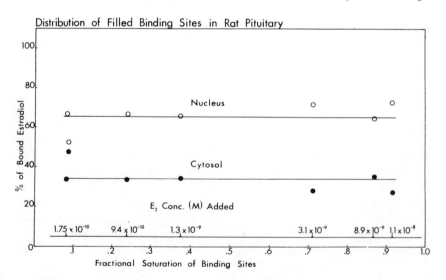

Fig. 1. *The distribution of the estrogen receptor between the cytoplasm and nucleus in the immature rat pituitary, as a function of the concentration of estradiol present during* in vitro *incubation. For experimental details see the methods section. Note that the distribution remains constant at all levels of saturation, from 8 to 90%, of the binding sites.*

tion over a sucrose density gradient permits the determination of the relative amounts of each form present. This type of experiment was performed using different concentrations of estradiol for the initial incubation. The results (see Figure 2) are in agreement with data from direct measurement of receptors and again show that the receptor distribution is independent of the number of receptors involved.

The higher proportion of the estrogen-receptor complex in the nucleus has, as stated in the introduction, been interpreted as a response to the presence of high affinity "acceptors" in the nucleus. If a limited number of acceptors were the determining factor in the nuclear localization, some evidence for saturation at the upper levels of filled receptor sites would be anticipated. In terms of the data we have presented one would have expected the percentage of nuclear localized receptor to decrease at higher concentrations of filled receptor sites and correspondingly the cytoplasmic forms to increase. The fact that the nuclear to cytoplasmic ratio for the distribution of receptor remains constant when measured by two different techniques and in two different tissues suggests that an acceptor hypothesis based on a limited number of high affinity acceptors is an over-simplification.

An alternative explanation for the intracellular distribution is that the estrogen receptor is located in accordance with the phase or polarity of the intracellular milieu. Using this rationale it would seem that the receptor must be restricted (by means discussed later) to the cytoplasm until it becomes bound to the hormone. It then may respond to the physicochemical environment in a passive manner.

A possible extension to this idea is offered by Horowitz and Moore (1974) who used the frog oocyte and autoradiographic techniques to study intracellular movements of solutes. Of relevance to the present discussion is their finding that radioactive inulin (an inert carbohydrate, m.w. 5,500), injected into the vegatative pole of the oocyte, concentrated within the nucleus, such that the nuclear to cytoplasmic ratio of the radioactivity on a grain density basis was 7.45:1. This preponderance of inulin within the nucleus is partially due to the amount of water present in the nucleus (1.9 times as much as that present in an equal volume of cytoplasm). To correct for this difference Horowitz and Moore expressed their results on a water basis but they still found more inulin in the nucleus with the nuclear to cytoplasmic ratio now 4:0. In an earlier study using radioactive sucrose (m.w. 342) these authors found that the nuclear to cytoplasmic ratio of injected sucrose expressed on a water basis was 1.6:1. To explain this difference of distribution between sucrose and inulin Horowitz and Moore hypothesized that not all of the water present in the cell may be available to solutes because of the existence of gels of macromolecular cellular constituents. These gel structures could conceivably act as molecular sieves preventing the free

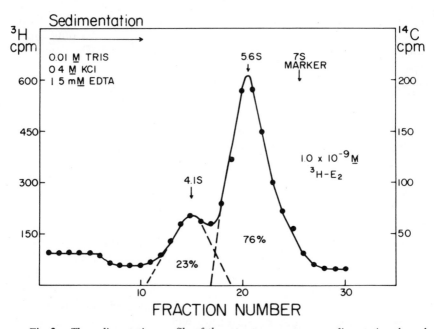

Fig. 2a. *The sedimentation profile of the estrogen receptor on sedimentation through a sucrose density gradient after extraction of whole uterine tissue with 0.4 M KCl. The uteri had previously been incubated with tritiated estradiol as described in the methods section.*

access of solutes to all "non-structured" areas of the cell. In a manner analogous to molecular sieves used in the laboratory, the excluded solutes would be distributed in accordance with the volume of the solution to which it has free access. All that is then required is that a higher percentage of the total water of the nucleus is actually available to inulin than that of the cytoplasm for a satisfactory explanation of the unequal distribution between the two compartments to be arrived at. The comparison of the distribution of radioactive glycerol, sucrose and inulin throughout the oocyte, the increase in the extent of exclusion from the cytoplasm of solutes as they increase in molecular weight and the diffusion pattern at the point of injection led Horowitz and Moore to conclude that an intracellular molecular sieve does, in fact, operate.

In their paper, Horowitz and Moore suggest that the estrogen-receptor may respond to these water-excluding forces. We find this an interesting possibility and feel that consideration of such phenomena is appropriate when discussing the translocation of the estrogen-receptor complex to the nucleus. Furthermore, a combination of this theory and the influence of differences in the phases of the nucleus and the cytoplasm seem to us to provide sufficient

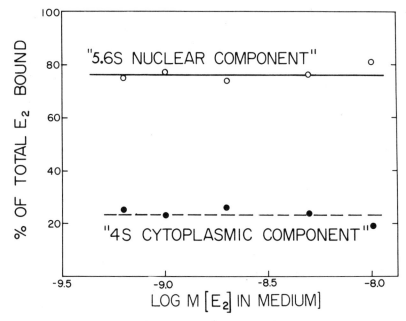

Fig. 2b. *A plot of data from a series of experiments similar to that shown in Figure 2a. The percentages of receptor present in the cytoplasmic 4S peak and the nuclear 5.6S peak was calculated from gradients similar to that shown in Figure 2a. These percentages were then plotted against the concentration of estradiol present during the tissue incubation. Note that the ratio of the cytoplasmic and nuclear forms of the receptor remains constant at all concentrations of estrogen used.*

explanation for translocation. A new model incorporating these ideas is shown in Figure 3. In it, the receptor, in the absence of estradiol is almost entirely in the cytoplasm, as has been observed frequently. Two possible explanations for this restriction are (a) the receptor in the absence of hormone is bound to some cytoplasmic component, or (b) the size of the receptor does not permit entry to the nucleus (studies of Gurdon (1970) and Paine and Feldherr (1972) have suggested that proteins of molecular weight greater than 60,000 can enter nuclei only to a very limited extent and extremely slowly). When estrogen enters the cytoplasm and binds to the receptor, the ability of the receptor to move within the cell is realized. We cannot, however, say whether this is a result of release of bound receptor or reduction of receptor size which would allow entry through nuclear pores. In this context it is difficult to accommodate the many findings (Jensen and DeSombre, 1972) concerning receptor transformation from a 4S form to a 5S form (i.e. from an entity with a molecular weight 76,000 to one of 132,000 (Notides and Nielsen, 1974)) prior to its rapid entrance through the nuclear membrane into the

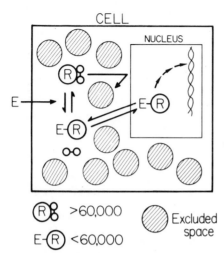

Fig. 3. *The cytoplasmic exclusion model for the translocation of the estrogen-receptor (E-R) into the nucleus. Details of the model are described in the text. E = estrogen, R = receptor.*

nucleus. Recent results from Siiteri et al. (1973) that transformation of receptor from 4S to 5S occurs within the nucleus may have a bearing on this point. Equally pertinent may be the suggestion by Jackson and Chalkley (1974) that the receptor does not, in fact, enter the nucleus but remains bound to the nuclear membrane in an artefactual manner and thus cannot be distinguished from a truly intranuclear localization. For the present, we choose to show the concentration of the estrogen-receptor within the nucleus to be a result of the exclusion phenomena described by Horowitz and Moore (1974). Translocation to the nucleus is therefore seen to function independently of the presence of acceptors and occurs in a manner readily reconcilable with a constant distribution of receptor between the cytoplasm and the nucleus irrespective of the number of estrogen-receptor entities involved. This model confers a position of great importance on the cellular architecture as it predisposes the ultimate location of any molecule in solution within the cell waters.

Nothing in the previous discussion argues against the existence of acceptors for the receptor on chromatin and indeed one could readily postulate that on entering the nucleus (as a consequence of exclusion phenomena) the receptor, as a secondary occurrence, binds to a high affinity acceptor. However, having discussed one of the pointers to the existence of acceptors (i.e. an uneven distribution of the estrogen-receptor complex within the cell) we must now re-examine two other arguments advanced in support of their existence (see introduction).

The first of these is the requirement of a high concentration of KCl to remove the receptor from the nucleus. That the receptor binds readily and with great tenacity to a variety of inert materials (e.g. glass beads) has been well recognized. Consequently, artifactual binding of the receptor to a nuclear component could be generated easily by homogenization of the tissue and makes the observation of tight binding an unreliable basis for hypothesis.

The isolation of the "acceptor" from the nucleus would prove its existence conclusively. Unfortunately, there are presently several such claims and they are self contradicting in the sense that one describes the "acceptor" as a histone protein (Puca *et al.*, 1974), whereas another finds the "acceptor" to be a non-histone acidic protein (Spelsberg *et al.*, 1971). Obviously, the validity of both claims remains clouded as a consequence and one might reasonably ask if the designation of "acceptor" is being assigned merely to those nuclear constituents which bind the receptor with highest affinity under the particular experimental conditions employed. Because the receptor readily interacts artifactually with a wide variety of cellular and noncellular matrices, we feel that clear demonstration of the isolation of the putative "acceptor" has not been achieved.

Those who assume that "acceptors" do in fact exist define them as having a high affinity for the receptor and being very few in number. These properties have been attributed not as a consequence of experimental observations but as postulates extrapolated largely from other studies on control of gene expression, in particular from studies on the lac operon of *E. coli* where the controlling repressor binds with very high affinity to a very limited number of sites on the DNA of the bacterium (Gilbert and Muller-Hill, 1967; Riggs *et al.*, 1970). It should be noted that in fact the analogy is inaccurate because the estrogen-receptor plays the role of a metabolic activator and as such might not have properties akin to the repressor protein. Apart from such theoretical considerations, it is difficult to fit experimental data to the concept of a very limited number of "acceptor" sites. It has been calculated that there are approximately 20,000 receptors in an average rat uterine cell under physiological conditions and at proestrous approximately 50% or 10,000 estrogen-receptor complexes are present in the nucleus. In a recent study, Chamness *et al.* (1974) showed that in a cell-free system no saturation of binding of estrogen-receptor complex to nuclei was observed at concentrations up to 3 picomoles (approximately twice that present *in vivo*). Furthermore, they suggested that previous reports, which showed saturation of binding, were inaccurate due to the artifactual interference of non-receptor cytoplasmic proteins with the nuclear binding of the receptor (Chamness *et al.*, 1974). Another indication of the non-saturability of the nucleus in its retention of receptor comes from the studies of Higgins *et al.* (1973). They report that nuclei from cells preincubated with steroid (glucocorticoid) to move all of the cytosol receptor to the nucleus could bind added receptor in a cell free system as would nuclei not preincubated with steroid. With both types of nuclei the saturation curves *in vitro* appeared to be exactly the same.

They concluded, that receptor interaction with nuclei was different *in vivo* as compared to *in vitro* however they explained their results in terms of two classes of acceptor sites rather than the non-saturation of the "acceptor".

The acceptor hypothesis envisages the binding of the estrogen-receptor complex to a few high affinity acceptor sites in the nucleus. As a corollary one would predict that the response of the tissue would be saturated once a relatively few receptors entered the nucleus and bound to the acceptors. In the rat uterus a characteristic response to estrogen is the production of a specific protein (i.p. or induced protein). We have examined the relationship between nuclear bound estrogens and the rate of synthesis of this protein (Figure 4). These experiments were carried out in a completely *in vitro* system using intact tissue with protein synthesis and estrogen binding determined in parallel incubations. With three different estrogens the results indicate a direct proportionality between binding and response up to the limit of available estrogen binding sites ($\sim$ 1.5 p moles/uterus). The failure to see a decrease in response at higher levels of bound estrogen suggests again that no factor is rate limiting under the conditions used in this study. Anderson *et al.* (1975) have reported that *in vivo* there is a simple relationship between estrogen binding at 1 hr and uterine wet weight at 6 hrs after injection. Similarly, Tomkins has reported a linear relationship of glucocorticoid-receptor complex and response (Tomkins *et al.*, 1970). All of these results would support a direct relationship between the number of hormone-receptor complexes in the nucleus and the cellular response. Recalling that the highest concentration of estrogen-receptor present in our study represents approximately 20,000 receptors in the nucleus, it becomes very difficult to envisage a direct specific interaction of the receptor with an "acceptor" present in very few copies. The failure to detect saturation of response could be due to multiple copies of the gene being involved in the response. Analysis of this possibility offered by both Harris *et al.* (1973) and Palacios *et al.* (1973), however, showed that only 1 gene copy was involved in the production of ovalbumin, a protein made in large amounts in response to steroid hormones in the chick oviduct. In the absence of evidence to the contrary it seems probable that this will prove to be the case for other hormone-responsive genes also.

One is therefore faced with the situation that the properties attributed to the "acceptor" (i.e. high affinity and few in number) are at least difficult to reconcile with evidence obtained from experiments designed to test them. In addition, the results most frequently presented in support of the existence of acceptors of the above type are subject to alternative and equally plausible explanations.

While presenting these arguments against the presence of "acceptors" as usually conceived, we are nonetheless aware that the estrogen-receptor must interact with "something"(s) in the nucleus to elicit its response. By defini-

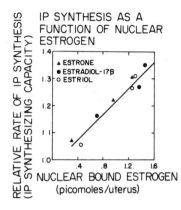

Fig. 4. *The relationship between estrogen receptor level in the nucleus and the relative rate of synthesis of I.P. (induced protein). The relative rates of I.P. synthesis and nuclear binding of estrogen at various concentrations of in vitro estrogen were determined in parallel flasks. The relative rate of I.P. synthesis was then plotted as a function of nuclear bound estrogen.*

tion, this "something" must be an acceptor; the question really is, what are the properties of the true acceptor? Observation of a free equilibrium between cytoplasmic and nuclear receptors leads to the possibility that the nuclear receptor is, for the most part, in solution in the nucleoplasm. The estrogen-receptor complex could then elicit its effect indirectly; e.g., by acting as an enzyme. We have examined this question, using conditions under which the equilibrium between the two compartments has been perturbed, and conclude the receptor in the nucleus is not freely in solution (Gannon and Gorski, to be published). The equilibria and lack of saturation data therefore indicate that the binding which occurs in the nucleus is of low affinity. A model similar to this proposal has recently been presented by Yamamoto and Alberts (1975). Studying the binding properties of the estradiol receptor for DNA they conclude that the interaction is of low affinity (Kd = $\approx 5 \times 10^{-4}$M) (Yamamoto, 1974) and is non-specific for the DNA base sequence (Yamamoto and Alberts, 1974). In the model which they have developed (Yamamoto and Alberts, 1975), the estrogen-receptor complex is translocated to the nucleus in response to the vast number of these low affinity binding sites. They also argue that the operator genes, analogous to the "acceptors" in this discussion, are masked by the low affinity binding but if detected would be few in number (between 1 and 10^3) and would have affinities for the estrogen receptor in the range of $10^{-8} - 10^{-10}$M. They propose that the equilibrium between high and low affinity sites prevents saturation of the high affinity sites.

While this model is very close to our ideas, we feel that it is incomplete. One obvious shortcoming is the preoccupation with DNA as the nuclear binding site. Clearly the existence of similar low affinity binding sites on chromatin would be anticipated and should be examined. More fundamentally, one finds in the model of Yamamoto and Alberts the same assumption that is made by those who advocate high affinity acceptors; translocation is due to the binding of the estrogen-receptor complex to chromatin sites. No data, other than circumstantial evidence, has been presented to justify this assertion. While we do not doubt that the ultimate location of the estrogen-receptor complex may be on chromatin, we question if the affinity of the estrogen-receptor complex for DNA (5×10^{-4}M for 5S, 6×10^{-3}M for the cytoplasmic 4S form) is any higher, or the number of such binding sites any greater than other low affinity interactions existing in the cytoplasm. The nature of these cytoplasmic components is a matter of conjecture, but one could point to the cytoplasmic 8S form of the estrogen receptor as an example of the type of protein-protein interaction which might be involved. The location of the estrogen-receptor complex in the cell would then become a rather delicate balance between two banks of low affinity sites. If this situation does exist, then we propose that the water exclusion phenomenon which we have discussed could become a crucial factor in shifting the bulk of the receptors from the cytoplasm to the nucleus. The existence of cytoplasmic interactions could also explain why a considerable percentage of the estrogen-receptor complexes remain in the cytoplasm even when an infinite binding potential exists in the nucleus.

In this paper we have discussed the translocation of the receptor as an event independent of nuclear binding. The separation of the movement of the receptor from the responses it subsequently promotes has permitted a new examination of the "acceptor" theory and of one of the assumptions inherent in the model of Yamamoto and Alberts. Finally, we have highlighted the cytoplasmic exclusion theory of Horowitz and Moore which, in addition to a distribution in response to phase considerations, seems to readily explain the translocation of the estrogen-receptor complex to the nucleus thereby raising questions about that particular role of chromatin binding.

ACKNOWLEDGEMENTS

The authors wish to acknowledge the expert help of Bobbi Maurer, Wendy Radtke and Christina Haak in the preparation of this manuscript. Recent work from our laboratory discussed in this paper was supported by the College of Agricultural and Life Sciences, University of Wisconsin, Madison and by research grants HD 08192 from the United States Public Health Service and Ford Foundation Grant No. 6300505A.

REFERENCES

Anderson, J. N., Peck, Jr., E. J. and Clark, J. H. (1975). *Endocrinology* **96**, 160–167.
Burton, K. A. (1956). *Biochem. J.* **62**, 315–317.
Chamness, G. C. and McGuire, W. L. (1972). *Biochemistry* **11**, 2466–2472.
Chamness, G. C., Jennings, A. W. and McGuire, W. L. (1974). *Biochemistry* **13**, 327–331.
Gilbert, W. and Muller-Hill, B. (1967). *Proc. Nat. Acad. Sci. U.S.A.* **58**, 2415–2421.

Gorski, J., Toft, D., Shyamala, G., Smith, D. and Notides, A. (1968). *Rec. Progr. Horm. Res.* **24**, 45–80.
Gschwendt, M. and Hamilton, T. H. (1972). *Biochem. J.* **128**, 611–616.
Gurdon, J. B. (1970). *Proc. R. Soc. Lond. (B) Biol. Sci.* **176**, 303–314.
Harris, S. E., Means, A. R., Mitchell, W. M. and O'Malley, B. W. (1973). *Proc. Nat. Acad. Sci. U.S.A.* **70**, 3776–3780.
Higgins, S. J., Rousseau, G. G., Baxter, J. D. and Tomkins, G. M. (1973). *Proc. Nat. Acad. Sci. U.S.A.* **70**, 3415–3418.
Horowitz, S. B. and Moore, L. C. (1974). *J. Cell Biol.* **60**, 405–415.
Jackson, V. and Chalkley, G. R. (1974). *J. Biol. Chem.* **249**, 1615–1627.
Jensen, E. V. and DeSombre, E. R. (1972). *Biochemistry* **41**, 203–230.
Jensen, E. V., DeSombre, E. R., Hurst, D. J., Kawashima, T. and Jungblut, P. W. (1967). *Arch. Anat. Microsc. Morphol. Exp.* **56**, Suppl. 3–4, 547–569.
Jensen, E. V., Suzuki, T., Kawashima, T., Stumpf, W. E., Jungblut, P. W. and DeSombre, E. R. (1968). *Proc. Nat. Acad. Sci. U.S.A.* **69**, 632–638.
Jensen, E. V., Numata, M., Brecher, P. I. and DeSombre, E. R. (1971). *In* "The Biochemistry of Steroid Hormone Action" (R.M.S. Smellie, ed.), pp. 133–159. Academic Press, New York.
Jungblut, P. W., Hätzel, I., DeSombre, E. R. and Jensen, E. V. (1967). *Colloq. Ges. Physiol. Chem.* **18**, 58–86.
Katzenellenbogen, B. and Gorski, J. (1972). *J. Biol. Chem.* **247**, 1299–1305.
Noteboom, W. D. and Gorski, J. (1963). *Proc. Nat. Acad. Sci. U.S.A.* **50**, 250–255.
Notides, A. C. and Nielsen, S. (1974). *J. Biol. Chem.* **249**, 1866–1873.
Paine, P. L. and Feldherr, C. M. (1972). *Exp. Cell Res.* **74**, 81–98.
Palacios, R., Sullivan, D., Summers, N. M., Kiely, M. L. and Schimke, R. T. (1973). *J. Biol. Chem.* **248**, 7530–7539.
Puca, G. A. and Bresciani, F. (1968). *Nature (Lond.)* **218**, 967–969.
Puca, G. A., Vincenzo, S. and Ernesto, W. (1974). *Proc. Nat. Acad. Sci. U.S.A.* **71**, 979–983.
Riggs, A. D., Bourgeois, S. and Cohn, M. (1970). *J. Mol. Biol.* **53**, 401–417.
Shyamala, G. and Gorski, J. (1969). *J. Biol. Chem.* **244**, 1097–1103.
Siiteri, P. K., Schwartz, R. E., Moriyama, I., Ashby, R., Linkie, D. and MacDonald, P. C. (1973). *In* "Receptors for Reproductive Hormones" (B. W. O'Malley and A. R. Means, eds.), pp. 97–112. Plenum Press, New York.
Spelsberg, T. C., Steggles, A. W. and O'Malley, B. W. (1971). *J. Biol. Chem.* **246**, 4188–4197.
Stancel, G. M., Leung, K. M. T. and Gorski, J. (1973a). *Biochemistry* **12**, 2130–2136.
Stancel, G. M., Leung, K. M. T. and Gorski, J. (1973b). *Biochemistry* **12**, 2137–2141.
Tomkins, G. M., Martin, Jr., D. W., Stellwagen, R. H., Baxter, J. D., Mamont, P. and Levinson, B. B. (1970). *Cold Spring Harbor Symp. Quant. Biol.* **35**, 635–640.
Williams, D. and Gorski, J. (1972). *Proc. Nat. Acad. Sci. U.S.A.* **69**, 3464–3468.
Williams, D. and Gorski, J. (1974). *Biochemistry* **13**, 5537–5547.
Yamamoto, K. R. (1974). *J. Biol. Chem.* **249**, 7068–7075.
Yamamoto, K. R. and Alberts, B. (1974). *J. Biol. Chem.* **249**, 7076–7086.
Yamamoto, K. R. and Alberts, B. (1975). *Cell* **4**, 301–310.

INVESTIGATION OF THE MECHANISMS AND CONSEQUENCES OF STEROID HORMONE ACTION ON VITELLOGENIN SYNTHESIS IN *XENOPUS LAEVIS*

Lawrence J. Wangh[1]
*National Institute for Medical Research, Mill Hill
London, N.W.7, England*

Ruth F. Longthrone
*MRC Laboratory of Molecular Biology, Hills Road
Cambridge, CB2 2QH, England*

John Knowland
*Department of Anatomy, the Medical School, University Walk
Bristol BS8 1TD, England*

I. INTRODUCTION

Our understanding of the molecular mechanisms by which steroid hormones act on target tissues has now reached a level of great sophistication in several experimental systems. The existence of steroid-specific receptors within the cytoplasm of target cells, as suggested by Jensen and Jacobson (1962) and Toft and Gorski (1966), has been confirmed. The receptors have been isolated and are known to interact with the nuclear chromatin. Hormone activation triggers the synthesis of both new messenger RNA and new proteins (O'Malley et al., 1975), as well as initiating a great many other nuclear and cytoplasmic changes. Our knowledge of these molecular events has been extensively reviewed (O'Malley and Means 1974) and will be thoroughly discussed by other authors in this volume.

It is now possible to investigate the molecular elements and molecular mechanisms that control specific genes in the complex eukaryotic genome. One approach (see the work of Gilmour and that of O'Malley in this volume) involves the isolation of suspected control proteins and observation of their interaction with chromatin *in vitro*. As an alternative approach we hope to

---

[1] Present address: National Institute for Medical Research, Mill Hill, London, N.W.7, England.

isolate receptor proteins from a target tissue which synthesizes a specific protein in response to a steroid, and to inject these proteins into cells of a different tissue to test their ability to activate the target tissue gene in cells in which this gene is normally inactive. For this purpose we plan to use oocytes or embryonic cells of *Xenopus laevis* as the recipient cells. This paper describes the properties of a liver tissue system from *Xenopus* that responds to steroid hormones *in vitro* by synthesizing easily recognised proteins.

In the frog, as in other oviparous animals, the major proteins of egg yolk are derived from a precursor protein, vitellogenin (Bergink and Wallace, 1974; Clemens, 1974), which is synthesized in the liver of mature females, secreted into the bloodstream, and absorbed by growing oocytes (Wallace and Jared, 1969). *In vivo*, estradiol regulates vitellogenin synthesis in females, and injection of the hormone induces vitellogenin synthesis in male liver, which normally does not produce this protein (Wallace, 1967; Schjeide and Lai, 1970; Clemens *et al.*, 1975).

Vitellogenin exists in the serum as a dimer with a total molecular weight of 450,000, fifty thousand daltons being due to attached lipids and covalently bound carbohydrate residues. The protein contains a high percentage of serine in the region of the molecule which is later released as phosvitin, and most of the serine residues are phosphorylated (reviewed by Clemens, 1974). In the blood, vitellogenin binds the bile pigment biliverdin and calcium and both are carried by vitellogenin into the oocytes. Growing oocytes within a particular size range (Glass, 1970) absorb vitellogenin by micropinocytosis. The oocyte requires the innermost of its three follicular cell layers in order to actively take up vitellogenin (Wallace and Jared, 1969). Within the oocyte vitellogenin is cleaved at specific sites to release the yolk proteins, lipovitellin and phosvitin. Lipovitellin contains two pairs of peptide chains and each lipovitellin unit is associated with two phosvitin molecules (Bergink and Wallace, 1974). In the *Xenopus* oocyte the yolk protein complex is sequestered in the yolk platelets, the core of which are revealed by electron microscopy to be hexagonally packed crystalline arrays. The yolk platelets form a gradient within the egg, being densely packed at the vegetal pole and more dispersed at the animal pole (Wallace, 1972). During embryogenesis yolk is gradually catabolized to supply building materials for the growing embryo (Glass, 1970). The location and exhaustion of yolk supplies within particular regions of the embryo have been correlated with embryonic differentiation (Slaughter and Triplet, 1975).

Thus the physiology of egg yolk protein involves four different tissues: the liver, the blood, the oocytes and the embryo. The protein chemistry involves synthesis, post-synthetic modification by three types of molecules, secretion, selective binding and absorption, peptide cleavage at specific sites, sequestration and crystallization, and finally gradual utilization. Some of these properties are summarized visually in Figure 1.

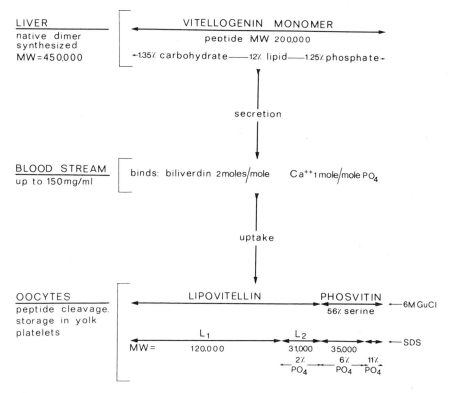

Fig. 1. See text for detailed explanation. Data taken from Clemens (1974) and Bergink and Wallace (1975).

The steroid hormone control of vitellogenin synthesis has been studied *in vivo* using the frog, the chicken, and a great many other egg-laying species (Schjeide and Lai, 1970). These efforts have established that estrogens (Redshaw *et al.*, 1969), but not other classes of steroids, both regulate vitellogenin synthesis levels in females and induce the synthesis of this protein in the liver of male animals. The earliest detected occurrence of yolk proteins in the serum following *in vivo* induction is 3-4 hours in the rooster (Bergink *et al.*, 1973), 9-12 hours in the frog (Dolphin *et al.*, 1971), and 24 hours in the fish (Plack *et al.*, 1971).

Explanted tissue from estradiol-stimulated animals has also been used to characterize the nature of the response to estradiol. Induced synthesis of yolk proteins can be detected during a twenty-four hour labelling period *in vitro* in tissue removed from animals three hours after estradiol treatment *in vivo* (Clemens *et al.*, 1975). Vitellogenin is the only phosphoprotein secreted by explants of vitellogenic liver (Wallace and Jared, 1969). These *in vivo* and tissue explant experiments have been thoroughly reviewed, (Schjeide and Lai, 1970; Clemens, 1974).

Neither *in vivo* nor tissue explant experiments prove that estradiol acts directly on the liver to induce and regulate vitellogenin synthesis. In fact, previously reported attempts to induce vitellogenin synthesis by treatment of liver *in vitro* have failed (Heald and McLachlan, 1965; Rudack and Wallace, 1968; Wallace and Jared, 1969; Plack and Fraser, 1971; Clemens *et al.*, 1975). Callard *et al.*, (1972) suggested the need for pituitary growth hormone. The relatively long delay before vitellogenin is first detected *in vivo* and in explants, and the high doses of estradiol administered (up to 1 mg/frog) make it imperative to prove that estradiol acts directly on the liver *in vitro*.

In vitro induction of the appropriate target tissue response has been accomplished in several other highly characterized steroid responsive tissue systems. O'Malley (1967) showed that pieces of immature chick oviduct from estrogen primed animals respond to progesterone treatment *in vitro* by increased avidin synthesis. Similarly, cells derived from chick oviduct and grown in culture for six weeks also show increased avidin and ovalbumin synthesis following treatment with progesterone and diethystilbestrol (O'Malley and Kohler, 1967). Cells from mature rat liver, immature rat liver and rat hepatoma cells have all been shown to respond to *in vitro* treatment with glucocorticoids by the increased synthesis of tyrosine amino transferase and other enzymes (reviewed by Wicks, 1974). Isolated rat uterus tissue responds within twenty minutes of *in vitro* addition of estradiol by the synthesis of the "IP" protein (Katzenellenbogen and Gorski, 1972). The polytene chromosomes of isolated diptera salivary glands undergo characteristic changes in puff pattern within fifteen minutes of the addition of ecdysone (Ashburner, 1972). Isolated oocytes of *Xenopus* and other frogs undergo maturation into eggs in the presence of progesterone (Jacobelli *et al.*, 1974). Finally, human lymphocytes maintained in culture rapidly begin the process of regulated cell death following the addition of glucocorticoids (Horibata and Harris, 1970).

Thus, by analogy with a large number of other tissues under steroid hormone control it is reasonable to expect that frog liver will rapidly respond to low doses of estradiol added *in vitro* if estradiol acts directly upon liver cells. Our first concern therefore was to establish culture conditions under which liver pieces taken from untreated animals could be induced to synthesize vitellogenin following treatment with estradiol *in vitro*.

II. MATERIALS AND METHODS

A. *Culture Media and Preparation of Liver Cultures.*

The culture medium (50%-MEM) was a 50% dilution of Eagle's Minimal Essential Medium with Earle's salts (Gibco Bio-Cult, Glasgow, Scotland) and contained a final concentration of 20 mM HEPES buffer pH 7.5, 2mM glutamine, 50 units/ml penicillin, and 50 μg/ml streptomycin. A similar medium (50%-MEMmet⁻) contained no methionine. Livers (0.5-2.5g) were taken from small, sexually mature frogs (13-45g) under sterile conditions into ice-cold 50%-MEM, chopped with scissors and rinsed thoroughly. About 60 mg of tissue (3-5 pieces) were placed in each well of a Linbro FB-16-24-TC culture dish and covered with 1 ml of 50%-MEM. Larger amounts of minced liver were maintained in suitable plastic petri dishes at a ratio of 60 mg tissue per ml of medium. Cultures were incubated in air at 25°C on a rocking table (4 cycles/min). The medium was changed daily. Estradiol-17β, progesterone, testosterone, and dexamethasone (all from Sigma) were added to 1 ml cultures in 5 μl of propylene glycol; control cultures received propylene glycol only.

B. Labeling of Liver Proteins with [^{35}S] Methionine.

The liver pieces were rinsed twice in 1-2 ml of 50%-MEMmet⁻ at 25°C and were then incubated for 6 hours at 25°C in 0.5 or 1 ml of medium containing 20 μCi/ml of [^{35}S]methionine (Amersham) at a suitable specific activity (see Results). In other experiments using labeling times of less than 6 hours tissue was incubated at 120 mg/ml in [^{35}S]methionine at 80 μCi/ml. At the end of the incubation period the culture was centrifuged at 2,000xg for 5 min at 4°C and the medium was removed and frozen at −20°C. The tissue was washed 3 times with 100 mM NH_4HCO_3, 5 mM $NaHSO_3$, 25 mM Tris-HCl, pH 7.0, and frozen. It was homogenized at 0°C in 0.5 ml of the same buffer containing 0.1% 2-mercaptoethanol and 150 μg/ml phenylmethylsulfonylfluoride in a microchamber Sorvall Omnimixer at top speed for 1 min. After taking aliquots to measure incorporation of isotope, the homogenate was centrifuged at 15,000xg for 20 min at 4°C and an aliquot of the supernatant was analysed by SDS gel electrophoresis. This centrifugation removed particulate material without altering the pattern of radioactive bands seen in whole homogenates. An additional aliquot of the supernatant was taken to determine incorporation of isotope.

C. SDS Gel Electrophoresis

10-90 μl of each culture medium were boiled with 10-50 μl of SDS gel sample solvent to give a final concentration of 2% SDS, 1% 2-mercaptoethanol, 10% glycerol, 0.001% bromophenol blue, 0.0625 M Tris, 10 mM Na_2EDTA, pH 6.8, and analysed by electrophoresis on 7.2% or 9% acrylamide slab gels as described by Knowland (1974). 50-100 μl of each homogenate supernatant were lyophilized, boiled in SDS gel sample mix of the final concentrations given above, and electrophoresed on 9% acrylamide slab gels. Each gel was fixed for 1 hour in 45% methanol, 7% acetic acid, and radioactive bands were located by fluorography (Bonner and Laskey, 1974) using hypersensitized film (Laskey and Mills, 1975). When [^{32}P]phosphate was used as the radioactive label the dye front containing free radioactivity was cut away before fixing the gel, and the radioactive bands were detected by autoradiography of the dried gel.

III. RESULTS AND DISCUSSION

A. Choice of $[^{35}S]$ methionine Specific Activity and Conditions for Tissue Culture.

The minimum requirement of any tissue culture system is that the cells remain viable and that their metabolic activity is detectable. We have used incorporation of $[^{35}S]$ methionine into protein as a criterion of cell health in our liver cultures. The culture conditions utilized may be compared to those used in the past in an effort to understand why previous attempts to detect induction of vitellogenin synthesis *in vitro* have failed.

When protein synthesis is monitored by incorporation of a radioactive amino acid, the concentration of non-radioactive molecules must be high enough to allow protein synthesis to proceed, without diluting the isotopic precursor so much that incorporation of radioactivity becomes inconveniently low. Figure 2 shows the effect of varying the methionine concentration on the uptake of methionine from the medium and its incorporation into protein synthesized by female liver during a 6 hour period. Uptake into the tissue and incorporation into both exported and non-exported proteins increase as the methionine concentration in the medium rises. On a linear scale the level of acid-soluble methionine within the tissue increases in direct proportion to the methionine concentration of the medium. The number of acid-insoluble counts in the medium is limited in 50%-MEMmet$^-$ due to low rates of protein synthesis, and is low in methionine-rich medium due to dilution of the isotope. The results suggest that a mixture of 9 volumes of 50%-MEMmet$^-$ and 1 volume of 50%-MEM, "9+1 medium", composes a suitable medium for active protein synthesis and experimentally convenient levels of isotope incorporation. For experiments involving labeling times as short as 0.5 hour we have used up to 40 μCi of $[^{35}S]$ methionine in 0.5 ml of 9+1 medium.

The viability of liver cultures maintained for up to two weeks has been compared to newly established cultures on the basis of $[^{35}S]$ methionine incorporation. Provided that the culture medium was changed daily, no difference was found in the level of isotope incorporation into protein in long term as compared to short term cultures. Experiments similar to those in Figure 2 showed that neither 5% female *Xenopus* plasma nor 5% male human serum had any effect on the rate of protein synthesis in liver cultures. The continued viability of long term cultures, maintained in the absence of added serum or plasma, also argues that these complex factors are not necessary. For this reason, and to avoid contamination with steroids present in plasma and serum, neither plasma nor serum was used in subsequent experiments.

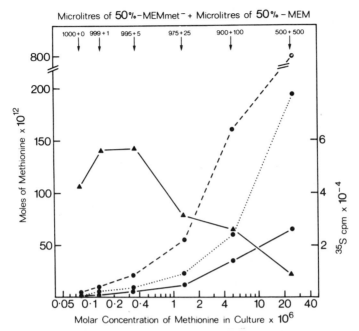

Fig. 2. Each culture of female liver contained 1 ml of medium prepared by mixing the volumes of 50%-MEMmet⁻ and 50%-MEM given at the top of the figure, and was incubated with 20 μCi [^{35}S]methionine (sp. act. 250 Ci/mMole, Amersham) for 6 hr at 25°C. Methionine uptake into tissue was determined from duplicate 20 μl aliquots of whole tissue homogenates dried on a GF/A filter (Whatman). Incorporation into protein was determined from duplicate aliquots, 100 μl for media and 20 μl for whole homogenates, precipitated with 10% TCA-5mM methionine, incubated 15 min at 90°C, collected and washed on filters, dried and counted at 80% efficiency. Results expressed as moles of methionine (X10^{12}): (● - - - - ●) total uptake into whole tissue homogenates; (● . . . . . . ●) acid-insoluble incorporation into whole tissue proteins; (●————●) acid-insoluble incorporation into secreted proteins; (▲————▲) ^{35}S cpm (X10^{-4}) recovered as acid-insoluble secreted protein in 100 μl aliquots of media.

B. Induction of Vitellogenin Synthesis by Estradiol-17β In Vitro.

Vitellogenin is secreted from liver cells soon after synthesis (Merry et al., 1973; Schirm et al., 1973). We find that tissue homogenates of labeled cultures contain little or no radioactive vitellogenin, even when large amounts are found in the medium. The fact that virtually all vitellogenin is exported greatly simplifies the detection and analysis of this protein.

The pattern of radioactive proteins found in the media from cultures of male or female liver in the presence or absence of estradiol-17β is shown in Figure 3. The results are typical of the effect of continuous hormone treatment *in vitro* for a period of several days. Male liver does not normally synthesize vitellogenin (track A), but is induced to do so by treatment with

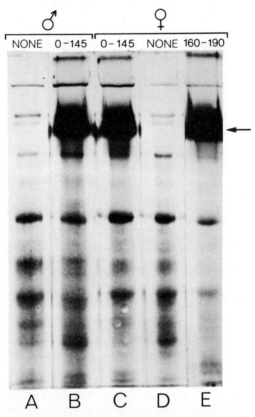

Fig. 3. *Radioactive protein bands in samples of media analysed on 7.2% acrylamide SDS-gel. Cultures A-D of male and female liver were maintained for 6 days in 50%-MEM without estradiol or with estradiol at 500 ng/ml (1.84 × 10^{-6}M). Culture E; was maintained for 7 days in 50%-MEM without estradiol and then for 30 hrs in 50%-MEM with added hormone. Cultures were labeled in 9+1 medium with or without added hormone as appropriate using [^{35}S]methionine as described (Fig. 1). The arrow on right indicates the position of the vitellogenin band.*

estradiol *in vitro* (track B). Other exported proteins characteristic of normal male liver continue to be synthesized and secreted in the presence of estradiol. Newly-established cultures of female liver synthesize vitellogenin (Wallace and Jared, 1969) and female liver cultured for several days in the presence of estradiol continues to synthesize large amounts of vitellogenin (track C), in addition to the other secreted proteins that exist in male liver cultures. In the absence of estradiol, vitellogenin synthesis in female liver cultures gradually declines, and by the sixth day without estradiol female liver no longer synthesizes vitellogenin, (track D), although synthesis of the other secreted proteins continues. If estradiol is added to these nonvitellogenic female cultures, synthesis of vitellogenin resumes within 30 hours (track E). Subsequent experiments showed that pieces of male liver cultured without estradiol for a week are also induced to synthesize vitellogenin after adding hormone. As judged from the patterns of radioactive bands on one dimensional SDS gels, vitellogenin is the only protein in either media or homogenates whose synthesis is induced by estradiol.

Approximately 10% (v/w) of freshly-excised liver is due to trapped serum (Wallace and Jared, 1969). Our culture media contain no serum or plasma. Thus, after one week in culture, with daily changes of the medium, all residual serum factors not synthesized by the liver will probably have been washed away. Comparison of the amount of hemoglobin in the homogenate supernatant of freshly excised liver and liver pieces maintained in culture for seven days reveals that the cultured liver contains about one seventh of the original amount of hemoglobin. Hemoglobin within red blood cells trapped in the liver pieces will wash away more slowly than serum proteins and therefore gives a maximum estimate of the amount of possible trapped serum factors. The fact that both male and female liver, after seven days in culture, can still respond to estradiol strongly suggests that no other factors are needed to induce synthesis of vitellogenin. We conclude that vitellogenin synthesis can be induced in male liver by treatment with estradiol *in vitro*, that sustained vitellogenin synthesis in both normal female and induced male liver requires the continued presence of estradiol, and that the hormone plus liver cells constitute a complete system for induction and control of vitellogenin synthesis.

C. *Previous Attempts to Induce Vitellogenin Synthesis* In Vitro.

Previous attempts to demonstrate the induction of vitellogenin synthesis in male liver treated with estradiol *in vitro* have been without success (Heald and McLachlan, 1965; Rudack and Wallace, 1968; Wallace and Jared, 1969; Plack and Fraser, 1971; Wallace and Berkink, 1974; Clemens *et al.*, 1975). Liver taken from male animals pretreated with estradiol six months previously shows a low level of vitellogenin synthesis. Wallace and Berkink (1974) reported that *in vitro* addition of estradiol to liver from pretreated males induces a small increase in the level of vitellogenin synthesis after more than 40 hours in culture. Several factors may be responsible for the previous failures to induce vitellogenin synthesis in naive male liver *in vitro*.

Incubation in a medium lacking or limited in essential amino acids causes tissue metabolic activity to decrease rapidly and protein synthesis is reduced to very low levels. Plack and Fraser (1971), for instance, reported that the level of protein synthesis in slices of cod liver *in vitro* decreased from 100% on day 1 to 0.5% on day 4, cultures were maintained in a medium lacking amino acids. Figure 2 demonstrates that in synthetically active female liver *in vitro* the total number of moles of methionine incorporated into protein during a 6 hour period in 50%-MEMmet⁻ with high specific activity methionine (250 mCi/ mole) is about $\frac{1}{80}$th of the amount incorporated into protein in medium with about 60 times the methionine concentration (3.9 mCi/ Mole).

We have found that the method used to detect vitellogenin synthesis significantly influences the conclusion as to whether or not induction of vitellogenin synthesis has been achieved. The detection system used here, SDS gel electrophoresis followed by fluorography, is direct and has high resolution. It displays all of the radioactive proteins synthesized and it distinguishes between them on the basis of molecular weight. Low levels of radioactivity can be detected in individual bands regardless of the number of specific activity of other radioactive bands. Thus the method is virtually independent of the percentage of total radioactivity represented by a single band. The detection methods used in the past have often involved acid precipitation of radioactive proteins in the medium from control and estradiol-treated liver samples. The hope is that treated samples will show significantly more radioactive secreted protein counts than control samples. This approach assumes that the induced protein represents a significant percentage of the secreted protein and that the total amount of protein synthesis must rise in treated cultures, which is not necessarily the case. We have found, using acid precipitation of radioactive secreted proteins from control cultures and cultures of male liver treated with estradiol *in vitro*, that the treated cultures are substantially higher than controls only after 140 hours of induction. By the electrophoretic method, induced vitellogenin synthesis is detected in the same cultures within a few hours of the addition of hormone.

The choice of radioactive precursor used to label proteins greatly influences the ease of detection and the conclusions drawn from the experiment. We have chosen [^{35}S] methionine because it is neither synthesized nor readily metabolized by the liver, which makes it easy to control its concentration. Its relative percentage in vitellogenin and many other proteins is similar, and it is available cheaply at high specific activity. Vitellogenin is unusually rich in serine, with 115 residues/1000 residues (Clemens, 1974), and the phosvitin polypeptide portion of vitellogenin contains 56% serine. In experiments using radioactive serine as the amino-acid label (Clemens *et al.*, 1975), a correction should be made for the unusual composition of vitellogenin.

Estimates of the amount of vitellogenin synthesized relative to other proteins will also be affected by the length of the radioactive labeling period. In experiments involving long period of radioactive labeling the radioactive precursor will accumulate in those proteins which turn over slowly. In contrast to some intracellular proteins, vitellogenin, being a secreted protein, does not turn over, and therefore is disproportionately labeled during long radioactive pulses. Thus, Clemens et al., (1975) have estimated that vitellogenin accounts for 90% of the protein synthesized by explants of liver fully stimulated *in vivo* and incubated *in vitro* for 12-24 hrs with radioactive serine. In contrast, we estimate, using a 6 hour pulse of methionine, that vitellogenin is only a few percent of the total secreted protein synthesized by liver pieces fully stimulated *in vitro*, (Table I). Moreover, the number of moles of methionine incorporated into total secreted protein is only about one half the number of moles of methionine incorporated into tissue homogenate protein (Fig. 2).

A more careful examination of the quantitative aspects of vitellogenin synthesis now seems warranted since such data may have relevance to the mechanism of liver differentiation in response to estradiol and the number of liver cells involved.

Several investigators have previously demonstrated that vitellogenin synthesis declines over several days in liver explants maintained in culture in the absence of estradiol (Clemens et al., 1975). This fact is more difficult to determine *in vivo* since withdrawal of estradiol, by ovariectomy results in a net increase in the blood levels of vitellogenin (Wallace and Jared, 1968). Our system which demonstrates that vitellogenin synthesis is inducible *in vitro* also allows estradiol to be withdrawn easily, and shows that sustained synthesis of vitellogenin by liver is dependent upon continuous exposure to estradiol.

TABLE I

Estradiol Concentration	% Total Counts in Vitellogenin Region*	Presence of Visible Vitellogenin band in Fluorograph
1. none	1.06	—
2. 2×10^{-10} M	1.52	—
3. 2×10^{-9} M	2.24	+
4. 2×10^{-8} M	6.23	+
5. 2×10^{-7} M	2.95	+

*Cultures of male liver were treated with estradiol as described in Figure 4. In order to determine total cpm of secreted proteins in the medium from each culture, samples were precipitated onto a filter using carrier bovine serum albumin in 15% trichloroacetic acid. Additional samples of each medium were analysed by SDS gel electrophoresis and fluorography. The location of vitellogenin was determined from the fluorographic image and that region was cut from the dried gel. Gel slices were dissolved in 500 μl of 30% H_2O_2 : 88% Ammonia :95:5 by incubation for 6 hrs at 37°C under a thin layer of light parafin oil. For scintillation counting each dissolved slice received 50 μl glacial acetic acid and 4 ml of Aquasol (New England Nuclear).

D. Estradiol Dose Response and Liver Response to other Hormones.

Additional criteria to prove that estradiol is the hormone responsible for regulating vitellogenin synthesis come from dose response experiments and experiments with non-estrogenic steroids. Redshaw *et al.*, (1969) injected estrogens, progesterone, testosterone and cortisone into frogs in order to determine the specificity of the *in vivo* response. Only estrogens induced synthesis of vitellogenin. We have incubated male liver for up to 10 days with each of the same four steroids. The liver cultures were maintained without added steroids for 4 days prior to addition of the hormone in order to wash out residual hormones and to allow any existing response to decline. The four steroids were added to separate cultures at four concentrations, from 2×10^{-10}M to 2×10^{-7}M in 10-fold steps. Every 3 days cultures were radioactively labeled and samples of media and homogenates were analysed as described in Methods.

The patterns in Figure 4 show the radioactive bands obtained from the media of cultures treated with estradiol, and homogenates of cultures treated with dexamethasone, for ten days. The only cultures that made vitellogenin were those containing 2×10^{-9} to 2×10^{-7}M estradiol-17β. No vitellogenin was found in either media or homogenates of cultures incubated with the other steroids or in estradiol at 2×10^{-10}M. Thus, the *in vitro* specificity of the liver response agrees with the estrogen dependence of the *in vivo* response. The lowest effective concentration of estradiol, about 2 nM, is similar to normal estrogen levels in the plasma of female vertebrates (Gallien and Chalumeau-LeFoulgoc, 1960; Marot and Ozon, 1973; Mikhail *et al.*, 1970) and close to the minimum effective concentration of steroids known to induce protein synthesis in other cell systems (Baxter and Tomkins, 1971; Sibley and Tomkins, 1974). Previous *in vivo* and tissue explant studies have involved administration of much higher levels of estradiol, 0.1 mg to 1.0 mg per frog, and cannot be used to determine the true sensitivity of the liver. These facts strongly suggest that estradiol-17β, rather than some minor contaminant, is the true inducing agent.

The direct visualization of radioactive proteins by electrophoresis and fluorography demonstrates that non-estrogenic steroids do not induce the synthesis of vitellogenin. Do they induce increased synthesis of any other protein? Synthesis of a second protein is induced by dexamethasone. As may be seen in Figure 4, this second protein has a lower molecular weight than vitellogenin and is found in the tissue homogenate only. Increased synthesis of this protein is induced in both male and female cultured liver by dexamethasone within hours of addition of the steroid at concentrations of 2×10^{-9}M or higher, and not by the other steroids tested. The physiological identity of this second protein is as yet unknown. The existence of two distinct hormone responses from a single tissue should be helpful in analysis of the mechanism of gene activation in this eukaryotic system.

Liver Protein Synthesis Following Estradiol or Dexamethasone Treatment

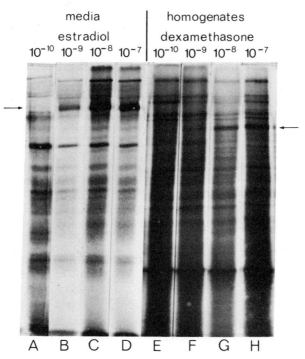

Fig. 4. *Radioactive protein bands in samples of media and homogenate supernatants on 9% acrylamide SDS-gel. Cultures of male liver were maintained for 4 days in 50%-MEM without added steroids. Incubation then continued for 10 days in 50%-MEM with estradiol-17β (A-D) or dexamethasone (E-H) added at concentrations of 2 × $10^{-10}M$ to 2 × $10^{-7}M$ in 10-fold steps. Cultures were radioactively labeled as in Fig. 3. The lefthand arrow indicates the position of vitellogenin (MW 200,000). A high level of synthesis of a different, predominantly non-secreted, protein (approximate MW 170,000) is induced by dexamethasone treatment* in vitro, *arrow on right.*

E. The Time Course of the Liver Response to Estradiol.

In each of the carefully studied cases, the initial response of the target tissue occurs very rapidly following the addition of the hormone. For instance, induced synthesis of the intracellular "IP" protein of rat uterus can be detected within 20 min of the addition of estradiol to tissue *in vitro* (Katzenellenbogen and Gorski, 1972). Dexamethasone stimulated tyrosine amino transferase activity within rat hepatoma cells has been detected 45 min after the addition of the hormone (Granner et al., 1970). Characteristic changes in the puffing pattern of salivary chromosomes occur within 15 min of the addition of ecdysone to the isolated glands (Asburner, 1972). The level of ovalbumin mRNA in chick oviduct cells has been quantitated using a probe of complementary copy DNA (cDNA). Newly synthesized ovalbumin mRNA molecules are detected 30 min after secondary stimulation with estradiol (O'Malley et al., 1975).

Using gel electrophoresis and fluorography we have tried to determine the minimum time following estradiol stimulation of male liver *in vitro* after which vitellogenin synthesis can be detected. Liver cultures were maintained for 7 days without hormone after which time estradiol was added along with [^{35}S]methionine at 80 μCi/ml. Samples of tissue (20 mg each) and incubation medium (150 μl) were taken at half-hour intervals and analysed as described previously. Radioactively labeled vitellogenin was detected in tissue homogenates and in the surrounding medium 2 hours after the addition of the hormone. Analysis of earlier time points is not yet complete. Additional experiments showed that vitellogenin synthesized *in vitro* can be radioactively labeled with [^{32}P]phosphate (Wallace and Jared, 1969; Wangh, unpublished results). Thus secreted vitellogenin molecules must have completed the process of post-translational phosphorylation. It is not known whether post-translational addition of carbohydrate and lipid groups has also occurred in this *in vitro* system.

The vitellogenin monomer is a large polypeptide which is synthesized at 25°C, is post-synthetically modified, and is secreted. Thus synthesis and processing of the molecule must require a significant amount of time even in fully-stimulated tissue. Clemens (1974) has estimated 120 min as the time required for the synthesis, processing, and secretion of vitellogenin in explants of fully stimulated *Xenopus* liver. This time is longer than the 20 min estimated for the same steps in chicken liver. The difference may reflect the fact that *Xenopus* cultures are maintained at 25°C while the temperature used for the chicken tissue was 37°C. We have thus far obtained a time of 2 hrs for the induction, synthesis, processing and secretion of vitellogenin. No vitellogenin mRNA is detected in the polysomes of unstimulated male liver (M.V. Berridge, personal communication). Thus if the time required for the synthesis, processing, and secretion of the polypeptide approaches 2 hrs the molecular mechanisms of estradiol induction, synthesis of vitellogenin mRNA and processing of the mRNA must be very rapid. It seems unlikely that induction of vitellogenin involves either the prior induction of some other protein or cell division.

F. Isolation of Vitellogenin Messenger RNA and Translation in Heterologous Systems.

PolyA-containing RNA has been prepared from the livers of estradiol treated animals and following heat denaturation the RNA was separated on sucrose gradients (M.V. Berridge, personal communication). Each fraction of the gradient was tested for its ability to support the synthesis of vitellogenin-like protein in a rabbit reticulocyte lysate protein synthesizing system. The vitellogenin-like proteins in the lysate were identified by immunoprecipitation followed by SDS gel electrophoresis and included a 210,000 MW polypeptide. The peak of vitellogenin messenger RNA activity in the sucrose gradient was at 29s. The number of nucleic acid bases in the vitellogenin message can be estimated to be about 5,000. Immunoprecipitation of polysomes labeled with orotic acid suggests that the message for vitellogenin is carried on about 25% of the total polysomes in the liver of a fully stimulated female frog (S. Farmer, personal communication). No vitellogenin mRNA is recovered from the polysomes of unstimulated male liver.

Total polysomal mRNA, as well as 29s mRNA, have been injected into frog oocytes. The injected RNA induces the synthesis of vitellogenin in the host cell (M. V. Berridge and C. Lane, personal communication). Thus the oocyte is capable of translating vitellogenin mRNA, in keeping with its ability to translate injected mRNA molecules for a large number of other proteins (Lane and Knowland, 1975).

G. Liver Cell Differentiation: Inflexible vs. Flexible Commitment.

The synthesis of vitellogenin by *Xenopus* liver in the presence of estradiol is an example of an adaptive and reversible response of a tissue to a specific environmental stimulus. A number of other examples of liver proteins whose synthesis is regulated by steroid hormones are listed in Table II. In those cases which have been examined closely, increased protein synthesis due to steroid induction is dependent upon the prior synthesis of mRNA molecules for each of the proteins. Thus in the absence of the inducer the particular gene in the liver appears to be inactive, but able to respond. On the other hand, there are genes in the adult liver which, as far as is known, are not responsive to environmental stimuli; for instance the genes for globin, keratin, and most of the "luxury" proteins characteristic of other differentiated tissues.

Vitellogenin and α_{2u} globulin (Table II) are normally sex-specific proteins whose synthesis in the liver of an animal of the opposite sex can be induced by the appropriate sex steroid. It is a striking fact that the liver of male frogs can respond to estradiol, despite the facts that, (1) males do not have ovaries, the normal source of estrogens; (2) males have very low blood levels of estrogens (Gallien and Chalumeau-Le Foulgoc, 1960); (3) the gonads of frogs only differentiate at metamorphosis (Blackler, 1972) well after the formation of the tadpole liver, and (4) the vitellogenin gene will normally never be called upon to function in the male.

The ability of liver to respond to estrogens and androgens clearly does not arise during tissue differentiation in a sex limited manner. Assuming that, as in other well documented examples of steroid action, frog liver contains a cytoplasmic receptor for estrogens the responsiveness of male liver means that the cells contain a constant level of the receptor even in the absence of the hormone.

TABLE II

Response	Hormone	Tissue	Reference
1. Vitellogenin synthesis	estrogens	*Xenopus* liver	Clemens (1974)
2. Gel band, MW 170,000	dexamethasone	*Xenopus* liver	Wangh and Knowland (1975)
3. α_{2u} globulin	androgens	rat liver	Sippel *et al.*, (1975)
4. Tyrosine aminotransferase, other liver enzymes	glucocorticoids	rat hepatoma cells rat liver	Tomkins *et al.*, (1969) Wicks (1974)
5. δ-Aminolevulinic acid synthetase	pregnene	rat liver	Edwards and Elliot (1975)
6. Nuclear non-histone protein MW 41,000 MW 26,000 and 20,000	cortisol estradiol	rat liver chick liver	Shelton and Allfrey (1970) Dierks-Ventling and Jost (1974)

It would be interesting to know if the livers of male and female tadpoles are capable of responding to estradiol by synthesizing vitellogenin. It is known that the liver of 15 day chick embryos responds to estradiol with the synthesis of serum phosphoproteins (Schjeide et al., 1960). By use of the sensitive analytical methods described in this paper it may be possible to investigate this question.

It has been extensively and elegantly demonstrated that the oviduct cells of immature chicks, and uterine cells of immature rats and calf can prematurely respond to estrogens by the synthesis of specific proteins and tissue differentiation (see the work of O'Malley, Jensen, and Gorski in this volume). Numerous other biological examples are known in which premature addition of a hormone triggers the precocious development of the adult response in the target tissue. Therefore, the structural and control genes related to each hormone must be potentially active in immature tissue despite the absence of the inducer. In this respect tissue differentiation is not dependent upon the age of the animal, as well as being independent of the sex of the animal.

It is useful to draw a conceptual distinction between the commitment which undifferentiated precursor cells make in becoming recognisable cells of a particular tissue type, and the commitment that already differentiated cells make in adaptively responding to a particular environmental stimulus. In general the commitment of precursor cells towards a particular line of differentiation is irreversible, (except under unusual experimental circumstances; see Harris, 1970; Gurdon, 1974). The adaptive response of differentiated cells is often reversible. These concepts have been extensively considered by others (Schjeide and de Vellis, 1970). It remains to be experimentally shown in detail how the molecular mechanisms of inflexible and flexible cell differentiation operate. Biological systems involving steroid hormone action, such as vitellogenin synthesis in estradiol stimulated frog liver, are amenable to detailed molecular analysis both at the level of inflexible and flexible cell differentiation.

ACKNOWLEDGEMENT

This work was supported by the Helen Hay Whitney Foundation of New York City (L.J.W.) and the Medical Research Council of Great Britain.

REFERENCES

Asburner, M. (1972). *In* "Developmental Studies on Giant Chromosomes" (W. Beermann, ed.), pp. 101–152. Springer-Verlag, Berlin.

Baxter, J. D. and Tomkins, G. M. (1971). *Proc. Nat. Acad. Sci. U.S.A.* **68**, 932–937.

Bergink, E. W., Kloosterboer, H. J., Gruber, M. and Ab, G. (1973). *Biochim. Biophys. Acta.* **294**, 497–506.

Bergink, E. W., and Wallace, R. A. (1974). *J. Biol. Chem.* **249**, 2897–2903.

Blackler, A. W. (1972). *In* "Oogenesis" (J. D. Biggers and A. W. Schuetz, eds.), pp. 321–338. University Park Press, Baltimore.

Bonner, W. M. and Laskey, R. A. (1974). *Eur. J. Biochem,* **46**, 83–88.
Clemens, M. J. (1974). *Progr. Biophys. Mol. Biol.* **28**, 69–107.
Clemens, M. J., Lofthouse, R. and Tata, J. R. (1975). *J. Biol. Chem.* **250**, 2213–2218.
Dierks-Ventling, C. and Jost, J-P. (1974). *Eur. J. Biochem.* **50**, 33–40.
Dolphin, P. J., Ansari, A. Q., Lazier, C. B., Munday, K. A. and Akhtar, M. (1971). *Biochem. J.* **124**, 751–758.
Edwards, A. M. and Elliot, W. H. (1975). *J. Biol. Chem.* **250**, 2750–2755.
Gallien, L. and Chalumeau-Le Foulgoc, M-T. (1960). *C.R.H. Acad. Sci.* **251**, 460–462.
Glass, L. E. (1970). *In* "Cell Differentiation" (O. A. Schjeide and J. de Vellis, eds.), pp. 201–223. Van Nostrand Reinhold Co., New York.
Granner, D. K., Thompson, E. B., and Tomkins, G. M. (1970). *J. Biol. Chem.* **245**, 1472–1478.
Gurdon, J. B. (1974). "The Control of Gene Expression in Animal Development" Clarendon Press, Oxford.
Harris, H. (1970). "Cell Fusion" Clarendon Press, Oxford.
Heald, P. J., McLachlan, P. M. (1965). *Biochem. J.* **94**, 32–39.
Horibata, K. and Harris, A. W. (1970). *Exptl. Cell Res.* **60**, 61–77.
Jacobelli, S., Hanocq, J., Baltus, E. and Brachet, J. (1974). *Differentiation* **2**, 129–135.
Jensen, E. V. and Jacobson, H. I., (1962). *Recent Progress in Hormone Research* **18**, 387–414.
Katzenellenbogen, B. S., and Gorski, J. (1972). *J. Biol. Chem.* **247**, 1299–1305.
Knowland, J. (1974). *Genetics* **78**, 383–394.
Lane, C. D., and Knowland, J. (1975) *In* "Biochemistry of Animal Development" (R. Weber, ed.), Vol. III pp. 145–181. Academic Press Inc., New York.
Laskey, R. A. and Mills, A. D. (1975). *Eur. J. Biochemistry*, in press.
Marot, J. and Ozon, R. (1973). *Comp. Biochem. Physiol.* **443**, 1103–1111.
Merry, A. H., Dolphin, P. J., Munday, K. A., and Akhtar, M. (1973). *Biochem. J.* **132**, 459–463.
Mikhail, G., Wu, C. H., Ferin, M. and Van de Wick, R. L. (1970). *In* "Karolinska Symposium on Research Methods in Reproductive Endocrinology" (E. Diczfalusy and A. Diczfalusy, eds.), 2nd Symposium, pp. 347–358. Stockholm.
O'Malley, B. W. (1967). *Biochemistry* **6**, 2546–2551.
O'Malley, B. W. and Kohler, P. O. (1967). *Proc. Nat. Acad. Sci. U.S.A.* **58**, 2359–2366.
O'Malley, B. W. and Means, A. R. (1974). *Science* **183**, 610–620.
O'Malley, B. W., Woo, S. L. C., Harris, S. E., Rosen, J. M., and Means, A. R. (1975). *J. Cell Physiol.* **85**, 343–356.
Plack, P. A., Pritchard, D. J., and Fraser, N. W. (1971). *Biochem. J.* **121**, 847–856.
Plack, P. A. and Fraser, N. W. (1971). *Biochem. J.* **121**, 857–862.
Redshaw, M. R., Follet, B. K. and Nicholls, T. J. (1969). *J. Endocrinol.* **43**, 47–53.
Rudack, D., Wallace, R. A. (1968). *Biochim, Biophys. Acta.* **155**, 299–301.
Schirm, J., Gruber, M., and Ab, G. (1973). *FEBS Lett.* **30**, 167–169.
Schjeide, O. A., Binz, S., and Ragan, N. (1960). *Growth* **24**, 401–410.
Schjeide, O. A. and De Vellis, J. (1970). "Cell Differentiation". Van Nostrand Reinhold Co. New York.
Schjeide, O. A. and Lai, G. A. B. (1970). *In* "Cell Differentiation" (O. A. Schjeide and J. De Vellis, eds.), pp. 447–475. Van Nostrand Reinhold Co., New York.
Shelton, K. R. and Allfrey, V. G. (1970). *Nature* **228**, 132–134.
Sibley, C. H. and Tomkins, G. M. (1974). *Cell* **2**, 213–220.
Sippel, A. E., Feigelson, P. and Roy, A. K. (1975). *Biochemistry* **14**, 825–829.
Slaughter, D. and Triplet, E. (1975). *Cell Differentiation* **4**, 11–22.

Toft, D. and Gorski, J. (1966). *Proc. Nat. Acad. Sci. U.S.A.* **55,** 1574–1581.
Tomkins, G. M., Gehehrter, T. D., Granner, D., Martin, D., Samuels, H. H. and Thompson, E. B. (1969). *Science* **166,** 1474–1480.
Wallace, R. A. (1967). *J. Cell Biol.* **35,** 137A.
Wallace, R. A. (1972) *In* "Oogenesis" (J. D. Biggers and A. W. Schuetz, eds.), pp. 339–360. University Park Press, Baltimore.
Wallace, R. A. and Jared, D. W. (1968). *Canad. J. Biochem.* **44,** 1647–1655.
Wallace, R. A. and Jared, D. W. (1969). *Dev. Biol.* **19,** 498–526.
Wallace, R. A. and Bergink, E. W. (1974). *Amer. Zool.* **14,** 1159–1175.
Wicks, W. D., (1974) In "MTP International Review of Science", (H. V. Rickenberg, ed.), Vol. 8, Biochemistry of Hormones, pp. 211–241. University Park Press, Baltimore.

III. Action of Plant Hormones

THE ROLE OF HORMONES IN THE INTEGRATION OF SEEDLING GROWTH

J. E. Varner

Department of Biology,
Washington University,
St. Louis, Missouri 63130

David Tuan-Hua Ho

MSU/AEC Plant Research Laboratory,
Michigan State University,
East Lansing, Michigan 48824

Heterotrophic growth of cereal seedlings (especially barley, wheat, oats, rice, maize and rye) depends upon mobilization of the proteins, carbohydrates and minerals stored in the endosperm (Fig. 1). This mobilization is brought about by the living cells of the outer layers of the endosperm — the aleurone layer (Yomo and Varner, 1971; Briggs, 1972). The easily observable agents of this mobilization are such hydrolases as α-amylase (EC 3.2.1.1.), protease (EC 3.4), β-glucanase (EC 3.2.1.6), ribonuclease (EC 3.1.4.) and acid phosphatase. The striking increases that occur in α-amylase (Yomo and Varner, 1971) and protease in embryoless half-seeds treated with gibberellic acid (GA_3) are due to *de novo* synthesis. The smaller increases in ribonuclease and in β-glucanase are also due to synthesis that is at least partly dependent on added GA_3 (Bennett and Chrispeels, 1972). Although the modest increase in acid phosphatase is apparently due to synthesis, the principal observable change in phosphatase is its release from the wall into the surrounding medium (Ashford and Jacobsen, 1974). This release is a part of the aleurone cell's response to GA_3 and is believed to result from the extensive cell wall degradation that follows the secretion of cell wall degrading enzymes — perhaps pentosanases (Briggs, 1963) because the principal polysaccharides of the wall are arabinoxylans (McNeil *et al.*, 1975). The metabolic behavior of the aleurone layer, and in particular the kinds of proteins being synthesized and secreted, is controlled by the growing embryo through the mediation of gibberellins. The response of the aleurone layers to GA_3 can be modified by the addition of abscisic acid (ABA) (Yomo and Varner, 1971). In addition the response of GA_3- treated layers to ABA can be modified by the addition of ethylene (Fig. 2). It is not yet clear whether the germinating seedling actually uses ABA and ethylene to modulate the aleurone layer's response to GA_3. However, one might argue

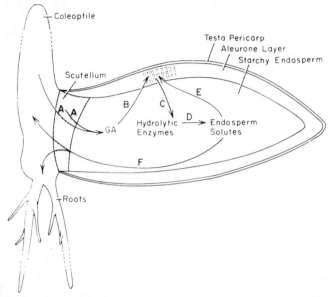

Fig. 1. *Diagrammatic representation of the relationship between gibberellic acid (GA) production, α-amylase production and solute accumulation in germinating barley seeds. GA produced by the coleoptile and scutellum (A) migrates into the aleurone layer (B) where hydrolytic enzyme synthesis and release is induced (C). These enzymes serve to hydrolyze the reserves in the endosperm (D), producing solutes which can inhibit further hydrolytic enzyme production (E), and function to nourish the growing embryo (F). From Jones and Armstrong, 1971.*

that the fact that the tissue has the ability to respond to ABA and ethylene favors such a modulation in the course of normal germination. The capacity of wheat aleurone layers to respond to GA_3 depends upon the diffusion of cytokinins from the starchy endosperm into the aleurone cells (Eastwood *et al.*, 1969).

The barley aleurone system was characterized in review by numerous photographs, light micrographs, electron micrographs, tables, and figures in 1971 (Yomo and Varner). For a review of the role of the aleurone layer in malting, see Palmer (1973); and for a more general review of the physiological role of gibberellins, see Jones (1973b). For the most part this present review deals with data published since 1971. For the uninitiated we show a schematic view of the barley seedling (Fig. 1) and the time course of α-amylase synthesis (Fig. 3) (protease synthesis precisely parallels α-amylase synthesis). The initiation of the rapid synthesis of α-amylase and protease at about eight hours is preceded by the proliferation of rough endoplasmic reticulum [as shown by electron micrographs (see Yomo and Varner, 1971)]. A

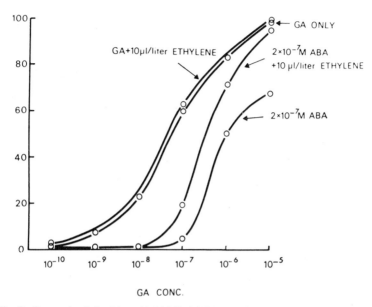

Fig. 2. *Reversal of abscisic acid (ABA) inhibition of amylase synthesis by increasing amounts of GA in the presence and absence of ethylene. All incubations were for 24 hr. The values given are for released α-amylase only and all are relative to the α-amylase production in the presence of 10^{-4} M GA_3 which has been given the value of 100. From Jacobsen, 1973.*

GA_3-dependent increase in the incorporation of labeled choline and of labeled phosphate into phospholipids probably reflects this membrane proliferation although a cautious interpretation is recommended (Firn and Kende, 1974). An even earlier effect, the GA_3-dependent increase in phosphorylcholine cytidyl transferase (EC 2.7.7.15) and in phosphorylcholine glyceride transferase (EC 2.7.8.2) as measured in cell-free extracts is detectable at two hours (Johnson and Kende, 1971) and even earlier as we shall see.

The principal approach to finding out how the aleurone cells respond to gibberellins has been to first describe the most easily observable responses, e.g. the massive secretion and release of hydrolases, the extensive cell wall degradation, the release of reducing sugars, the vacuolation of the cells and the degradation of the reserve proteins. These accumulated changes are, of course, the expression of a redirection of the cell's metabolism that occurs during the first few hours following treatment with GA_3. Such a redirection is in turn dependent on the presence, before the addition of GA_3, of a GA_3 receptor. We review first the late consequences of GA_3-treatment and work our way back in response time toward the earliest event — the formation of a gibberellin-receptor complex, an event not yet observed.

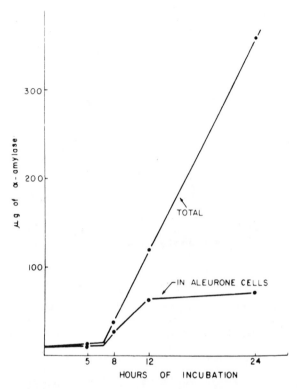

Fig. 3. *Time course of α-amylase synthesis by 10 aleurone layers incubated with 1 µM GA_3. Enzyme activity was measured in the medium surrounding the aleurone layers and in the supernatant of a 0.2 M NaCl extract of the aleurone layers. The term total refers to the sum of these two activities. From Chrispeels and Varner, 1967.*

Endosperm modification can begin only after the release of hydrolases from the aleurone layers into the starchy endosperm. Such release requires that the aleurone cells be bathed in a medium of a certain minimum ionic strength (Varner and Mense, 1972). The aleurone cells are the source of these ions (Table I). The ions are secreted from the aleurone cells in response to GA_3 (Fig. 4) and also during normal germination (Fig. 5). Enzyme release from the aleurone cells parallels, and probably requires degradation of the cell walls (Ashford and Jacobsen, 1974). Release of the hydrolases from the cell walls is, of course, preceded by secretion of the hydrolases from their site of synthesis into the periplasmic space and into the wall. This requires energy (Varner and Mense, 1972).

Aleurone layers, in midcourse production of α-amylase, secrete most of the newly synthesized protein within 60 minutes of the time of synthesis (Figs. 6 and 7). At the end of a 10 minute labeling period with tritiated

TABLE I.
Effect of Solutions of Ash of Aleurone Layers, Starchy Endosperm, and Liquified Endosperm on Release of α-Amylase.

Incubation medium	Steady state rate of α-amylase release
	μg/6 min
A. Standard	0.28
Distilled water	0.04
Solution of ashed aleurone layer	0.28
Solution of ashed aleurone layer diluted 6-fold	0.15
B. Standard	0.36
Standard without $CaCl_2$	0.08
Solution of ashed starchy endosperm	0.04
C. Standard	0.28
Distilled water	0.04
Solution of ashed liquid endosperm	0.31

Forty-eight barley seeds were germinated on moist sand for 5 days. The roots and shoots were removed, the liquified endosperm was collected with a pasteur pipette, and the aleurone layers were stripped off and rinsed in distilled water. The liquified endosperm and aleurone layers were ashed in separate crucibles, and the ash was dissolved in 12 ml 1 mM HCl and adjusted to pH 4.8. Starchy endosperm was recovered by dissection from 20 imbibed half-seeds and ashed, and the ash was dissolved in 8.0 ml mM HCl and adjusted to pH 5.0. All solutions were aerated before use. From Varner and Mense, 1972.

amino acids, the radioactivity is associated chiefly with the rough endoplasmic reticulum. From autoradiographs there is no evidence that the labeled protein is ever packaged in a vesicle as a part of the secretion process (Figs. 8-12; Chen and Jones, 1974a, 1974b), although it has been reported that α-amylase and protease are found in lysosomal-like vesicles in cell homogenates (Gibson and Paleg, 1972).

It is well established that the GA_3-dependent synthesis of hydrolases is prevented by RNA synthesis inhibitors added at the same time or shortly after the addition of GA_3 (Yomo and Varner, 1971). However, the synthesis of α-amylase is no longer susceptible to an RNA synthesis inhibitor added 10-12 hours or later after the addition of GA_3 (Fig. 13). Thus all kinds of RNA, including mRNA, required for α-amylase synthesis are stable *in vivo* for at least 14 hours. Because ABA added 12 hours after GA_3 inhibits α-amylase synthesis (Fig. 14), it is clear that ABA, directly or indirectly, can control α-amylase synthesis at the level of translation. The control may well be indirect because the addition of cordycepin immediately relieves the ABA inhibition of α-amylase synthesis (Fig. 14). This observation should be of great help in sorting out possible sites of action of ABA even though there is yet no

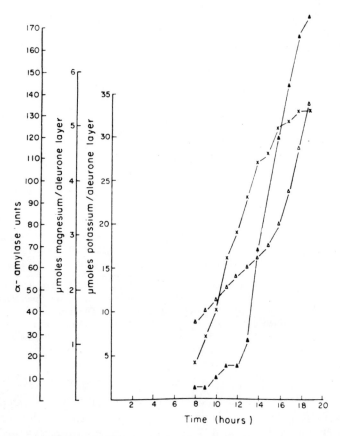

Fig. 4. *Time course of potassium (X), magnesium (△), and α-amylase (▲) release from 50 aleurone layers incubated in 25 ml of 5 μM GA_3 and 20 mM $CaCl_2$. From Jones, 1973a.*

direct evidence that ABA is involved in the control of α-amylase synthesis in the growing seedling. Is it possible that ethylene as it relieves the ABA inhibition of α-amylase synthesis (Fig. 2) also exerts its control at the level of α-amylase translation?

During the initial phase of the rapid accumulation of α-amylase and protease, there is a GA_3-dependent increase in the rate of synthesis of poly (A) RNA (Fig. 15) which can be presumed to be mRNA because it contains poly (A) and because it has an appropriate molecular weight (Fig. 16). It remains to be shown, by translation of the mRNA in a cell-free system, whether α-amylase mRNA synthesis parallels the GA_3- enhanced synthesis of poly (A) RNA. A direct assay of mRNA will also allow a resolution of the

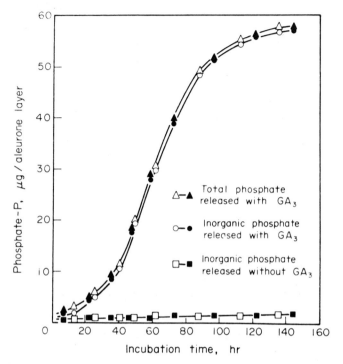

Fig. 5. *Release of total and inorganic phosphate from isolated aleurone layers incubated at 25°C with and without GA_3 (2 μg/ml). Open and closed symbols, two sets of incubations started 6 hr apart. Media were changed every 12 hr. The curve for organic phosphate release without GA_3 is coincident with the inorganic curve, and is not shown. Results are the means of six determinations. From Clutterbuck and Briggs, 1974.*

possibility that some or all of the α-amylase mRNA is present before the addition of GA_3 (Carlson, 1972).

Coincident with the increased rate of synthesis of poly (A) RNA, there is an increased rate of incorporation of $^{32}P_i$ into phospholipids (Fig. 17). The rate of labeling of phospholipids is proportional to the concentration of GA_3 added (Fig. 18) and at a constant concentration of GA_3 is inhibited proportionately to increasing concentrations of ABA (Fig. 19). It is clear, therefore, that the response of the cells to ABA does not lie solely at the level of translation of α-amylase.

Changes in the kinds of proteins being synthesized by GA_3-treated cells as compared with control aleurone cells are readily detected within two to four hours after the addition of GA_3 (Fig. 20). During midcourse α-amylase production about 40% of the total protein being synthesized is α-amylase (Fig. 21 and D. F. Flint, unpublished data) and α-amylase constitutes a major

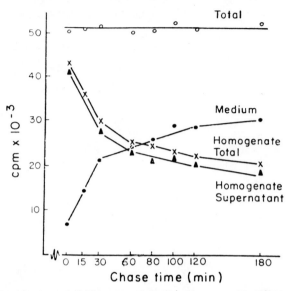

Fig. 6. *Distribution of [³H] leucine (2.5 µCi/ml; sp. act. 58 Ci/mM) in homogenate and medium fractions from 10 aleurone layers. Total, radioactivity recovered from all homogenate and medium fractions; homogenate total, radioactivity in supernatant and pellet fraction. The layers were pulse labeled for 10 min before chase began. From Chen and Jones, 1974a.*

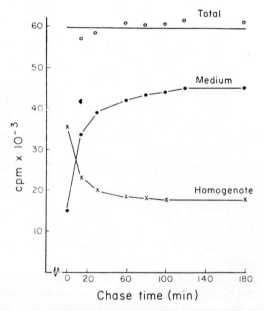

Fig. 7. *Distribution of [³H] tryptophan (2.5 µCi/ml; sp. act. 1.134 Ci/mM) in homogenate and medium fractions from 10 aleurone layers. Pulse and chase were carried out as described in Fig. 5. From Chen and Jones, 1974a.*

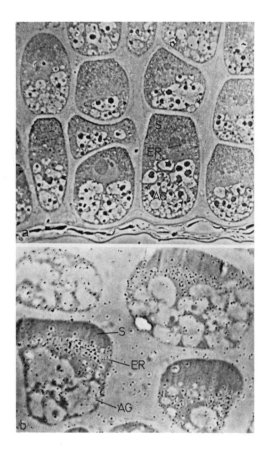

Fig. 8. (a) Light micrograph of control aleurone layer centrifuged at 105,000 g for 90 min. Each cell is clearly stratified into 3 zones; spherosomes (S), ER region and aleurone grains (AG). X 1,100. (b) Light microscope autoradiograph of stratified aleurone layers treated with GA_3 (5 μM) for 12 hr and pulsed with [^3H]leucine for 10 min. Note accumulation of silver grains over ER region of cell. X 1,500. From Chen and Jones, 1974b.

fraction of the total protein secreted or released (Jacobsen and Knox, 1974). Aleurone layers of barley contain large amounts of sucrose (30-40 μg/mg fresh weight). Treatment of the layers with GA_3 causes the release of sucrose and other sugars from the cell starting at two hours after the hormone (Fig. 22). Sucrose release probably does not require protein synthesis because it is not inhibited by high concentrations of osmotica such as polyethylene glycol and mannitol. If sucrose is released from the aleurone tissue during germination, it is presumably used by the growing embryo.

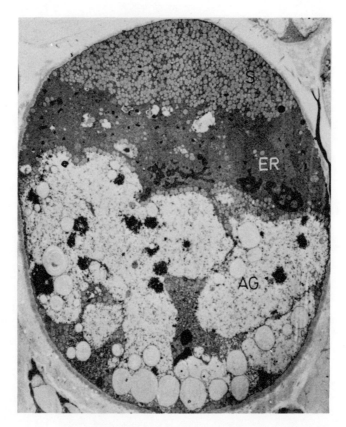

Fig. 9. High resolution autoradiograph of aleurone cell treated with GA_3 (5 μM) and $KBrO_3$ (5 mM) for 12 hr prior to pulsing with [^3H]leucine for 10 min. Silver grains are concentrated over the ER region of the stratified cell with few in the S and AG regions. X 3,600. From Chen and Jones, 1974b.

The GA_3-dependent increase of phosphorylcholine-cytidyl and phosphorylcholine-glyceride transferases becomes readily detectable within two hours of GA_3 treatment (Figs. 23 and 24). In wheat aleurone tissue the turnover rate of nucleotides, especially CTP, has been reported to be enhanced in 15 to 90 minutes after GA_3 (Collins et al., 1972). All these events seem to be a prerequisite of the later membrane proliferation because phosphotidylcholine (lecithin) is the major phospholipid in barley aleurone membranes (Koehler and Varner, 1973).

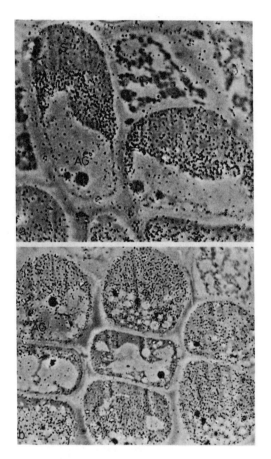

Fig. 10. Light microscope autoradiographs of stratified aleurone layers treated with GA_3 (5 μM) for 12 hrs and pulsed with [^3H] leucine for 10 min and chased with 10 mM carrier leucine for 60 min (a) and 120 min (b). (a) X 1,800; (b) X 1,400. From Chen and Jones, 1974b.

1. Choline + ATP $\xrightarrow{\text{choline kinase}}$ P-choline + ADP

2. P-choline + CTP $\xrightarrow{\text{phosphorylcholine-cytidyl transferase}}$ CDP-choline + PPi

3. CDP-choline + 1,2-diglyceride $\xrightarrow{\text{phosphorylcholine-glyceride transferase}}$ lecithin + CMP

However, the GA_3 enhanced membrane proliferation does not result in a net synthesis of phospholipids (Koehler and Varner, 1973). Since aleurone tissue has large lipid storage in spherosomes, membrane proliferation is probably the consequence of a hormone enhanced phospholipid turnover, i.e. from storage phospholipid to membrane phospholipid.

Apparently the increase in phosphorylcholine-glyceride transferase activity is due to some kind of activation because amino acid analogues, or cordycepin, do not prevent the increase (Ben-Tal and Varner, 1974) (Table II). If some kind of activation accounts for the early increase of phosphorylcholine-glyceride transferase activity following GA_3 treatment, one might expect to see some evidence for such activation *in vitro* by mixing homogenate of hormone-treated tissue with homogenate of control tissue. Such an effect is indeed observed within minutes of GA_3 addition (Ben-Tal and Varner, 1974; Ben-Tal, 1974) (Table III), but this activation can be prevented if ABA is present during the treatment of tissue with GA_3 (Ben-Tal and Varner, 1974) (Table III). Phosphorylcholine-glyceride transferase activity is inhibited by Ca^{++}. Therefore, the activity change of this enzyme could result from changes in concentrations of Ca^{++} or the concentrations of Ca^{++} complexing agents, such as citrate or phytate. Since aleurone tissue has a large storage of phytate,

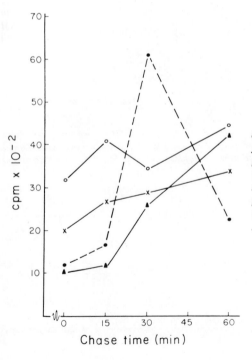

Fig. 11. *Distribution of radioactivity in pellet and supernatant fractions from 50 aleurone layers pulsed with labeled [^3H] leucine (2.5 µCi/ml) for 10 min.* ○——○, *low speed pellet;* ×——×, *mitochondrial pellet;* ▲——▲ *microsomal pellet;* ●——●, *supernatant. From Chen and Jones, 1974a.*

TABLE II.
Effect of GA_3 and Amino Acid Analogs on Phosphorylcholine Glyceride Transferase Activity in Barley Aleurone

Treatment during incubation	% of Control	Standard deviation
Control	100	
1 μM GA_3	161	±38.2
1 μM GA_3 and 7 mM amino acid analogs	167	±40.4

The enzyme was prepared from aleurone layers after 4 hr incubation of half-seeds. The controls were equal to 1229, 1022, and 1435 pmoles lecithin formed/hr 100 layers in three different experiments respectively, each one with two replicates. Because of large differences in the absolute pmoles of lecithin formed in different experiments the control in each experiment has been taken arbitrarily as 100%, and the treatments effect was compared on a percentage basis in each experiment separately, so that the standard deviations for the treatments would be calculated. From Ben-Tal and Varner, 1974.

TABLE III.
The Inhibitory Effect of Abscisic Acid on the Tissue Response to GA_3

Treatment No.	Treatment	Incubation time	Relative activity Observed	Expected	% Difference from expected
1	Control	4 hr	50		
2	GA_3	4 hr	72		
3	10^{-5} M ABA	4 hr	50		
4	GA_3 and 10^{-5} M ABA	4 hr	47		
	Mixture of 1 and 2		158	122	30
	Mixture of 2 and 3		162	122	33
	Mixture of 1 and 4		96	97	−1
	Mixture of 3 and 4		102	97	5
1	Control	8 hr	50		
2	GA_3	1 hr	84		
3	10^{-5} M ABA	8 hr	57		
4	GA_3 and 10^{-5} M ABA	1 hr	61		
	Mixture of 1 and 2		202	134	51
	Mixture of 2 and 3		199	141	41
	Mixture of 1 and 4		122	111	10
	Mixture of 3 and 4		118	118	0

Means of 3 different experiments. The expected values are the sums of the separate activities of the preparations involved in a given mixture. The amount of lecithin formed in the control tissue after 4 hr in the first part, and after 8 hr incubation in the second part was taken as 50% activity. From Ben-Tal, 1974.

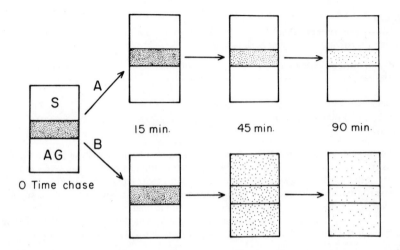

Fig. 12. *Diagrammatic representation of the distribution of silver grains in aleurone cells given a pulse of labeled amino acid and chased with excess carrier amino acid for 15, 45 and 90 min. Route A indicates the predicted distribution when labeled protein is released via a membrane bound vesicle, while route B shows the fate of label secreted directly from the cytoplasm without the participation of such vesicles. From Chen and Jones, 1974b.*

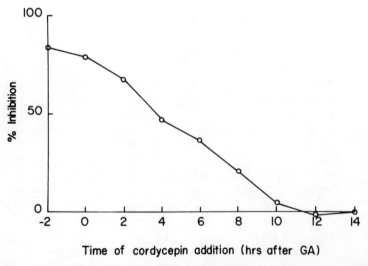

Fig. 13. *Effect of cordycepin on α-amylase activity. Cordycepin was added at different times after GA_3 as indicated, and aleurone layers were further incubated until 24 hr after addition of GA_3, when the activity of α-amylase was assayed. From Ho and Varner, 1974.*

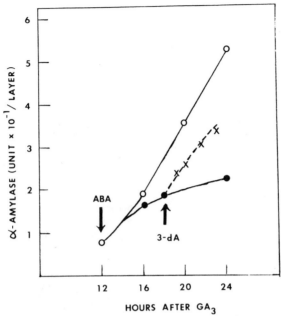

Fig. 14. *Effect of midcourse addition of ABA and cordycepin on the synthesis of α-amylase.* ○——○ GA_3 *only;* ●——● GA_3 *and ABA;* ×——× GA_3, *ABA, and cordycepin (3-dA); From Ho and Varner, in press.*

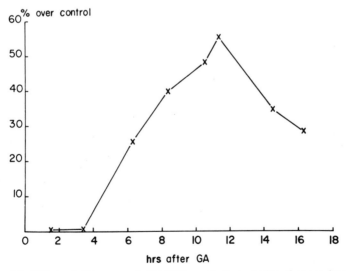

Fig. 15. *Effect of GA_3 on poly(A)-RNA synthesis in barley aleurone layers. From Ho and Varner, 1974.*

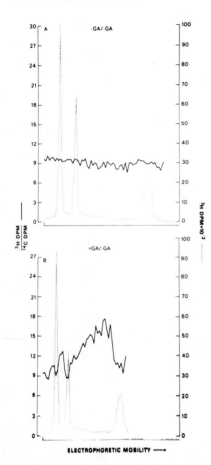

Fig. 16. *Polyacrylamide gel electrophoresis of double labeled RNA showing the effect of GA_3 on Poly(A)-RNA synthesis. (A) The $-GA_3/-GA_3$ RNA (solid ratio line) is RNA prepared from a mixture of aleurone layers, some of which were labeled with [^{3}H]adenosine only and some with [^{14}C]-adenosine only, both in the absence of GA_3. (B) Same as in A except some layers were labeled with [^{3}H]adenosine in the presence of GA_3. From Jacobsen and Zwar, Austral. J. Plant Physiol., in press.*

a redistribution of phytate or of calcium or an early release of calcium to the medium due to a change in membrane permeability could account for the activation of phosphorylcholine-glyceride transferase. The activity of membrane bound enzymes, such as phosphorylcholine-glyceride transferase, can be altered in the presence of a surfactant, such as lysolecithin which is one of the products of the hydrolysis of storage phospholipids. The activation of phosphorylcholine-diglyceride transferase begins within a very short time after hormone treatment; therefore, the mechanism of this activation process will certainly provide useful clues about the primary action of the hormone.

Another early response of aleurone cells is the GA_3-dependent susceptibility to the toxic effects of 1,10-phenanthralene (Goodwin and Carr, 1970).

Attempts have been made to look for the primary binding site of gibberellins in barley aleurone tissue but no helpful information is yet

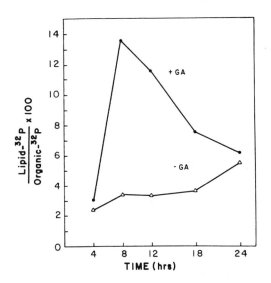

Fig. 17. *Time course of the rate of phospholipid labeling. Incorporation of ^{32}Pi into phospholipids is expressed as a percentage of the ^{32}Pi incorporation into organic phosphate. From Koehler and Varner, 1973.*

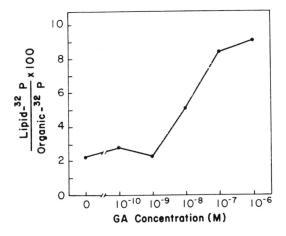

Fig. 18. *The response of phospholipid labeling to increasing concentrations of GA_3. Aleurone layers were incubated in solutions of the indicated GA_3 concentration for 10 hr. They were then pulse labeled with ^{32}Pi, and phospholipids were extracted. From Koehler and Varner, 1973.*

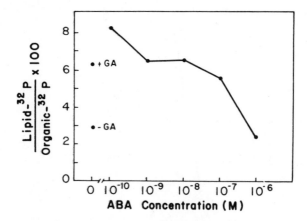

Fig. 19. *Progressive inhibition of phospholipid labeling by increasing concentrations of ABA. Aleurone layers were incubated for 8 hr with both 1.0 μM GA_3 and ABA at the indicated concentrations present from the start. The layers were then pulse labeled with ^{32}Pi, and phospholipids were extracted. From Koehler and Varner, 1973.*

available (Musgrave et al., 1972). Although aleurone cells do metabolize gibberellins (Musgrave et al., 1972) and ABA influences this metabolism (Nadeau, 1972), we feel that the overall rate of metabolism is insufficient to account for the rapid effects of ABA.

Cyclic AMP at high concentrations (about 5×10^{-3} M) has been reported to mimic certain effects of GA_3 in aleurone tissue, but the responses do not equal the full effects of GA_3 itself and may be pharmacological (see Lin, 1974). Careful investigation has ruled out the possibility that cyclic AMP is synthesized in response to GA_3 in barley aleurone layers (Keates, 1973). So far no convincing demonstration of adenyl cyclase has been obtained in plant tissues (see Lin, 1974) and although nucleotide phosphodiesterase is present in various plants, it has much less activity toward $3':5'$-cyclic AMP than toward $2':3'$-cyclic AMP which appears to be the product of RNA catabolism (Lin and Varner, 1972; Amrhein, 1974). Therefore, the occurrence of cyclic AMP in plant tissue is still in doubt and it seems unlikely that cyclic AMP mediates the action of GA_3 in aleurone layers.

There are two models that can account for the GA_3 enhanced synthesis of α-amylase (and other hydrolases).

1) The enhanced synthesis of α-amylase depends on the faster rate of formation (transcription and/or processing) of its specific mRNA. Membrane proliferation is a separate effect of GA_3.
2) The GA_3 enhanced membrane proliferation forms more sites for the synthesis of α-amylase. Thus, the amount of α-amylase specific mRNA

need not be increased in the presence of GA_3. A cell-free assay for the α-amylase specific mRNA should allow a decision between these alternatives.

ACKNOWLEDGEMENTS

This work was supported in part by the National Science Foundation (GB 39944) and by the United States Atomic Energy Commission (Contract No. E (11−1) 1338).

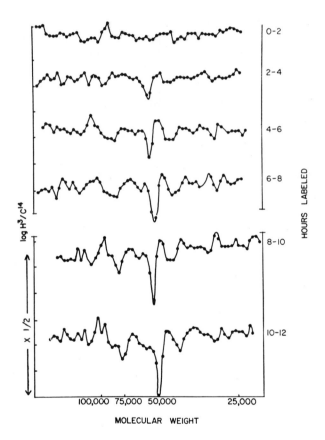

Fig. 20. *Labeling of the salt-soluble proteins of the control and GA_3-treated aleurone layers expressed as ^3H (control) to ^{14}C (GA_3-treated) ratios at incubation times up to 12 hr. From Varner, Flint and Mitra, 1975.*

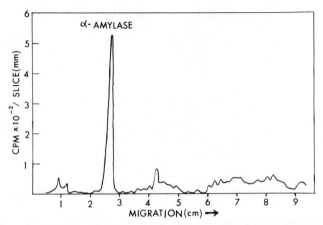

Fig. 21. *Profile of newly synthesized proteins in barley aleurone layers. Twenty-five aleurone layers were labeled with [^3H] leucine (20 µCi/ml incubation buffer) from 23 to 25 hr after GA_3. The layers were homogenized with 0.125 M Tris-HCl, pH 6.8, containing 4% SDS, 10% β-mercaptoethanol. After centrifugation at 12,000 g for 30 min, the protein profile in the supernatant was checked by SDS gel (10%) electrophoresis. No radioactivity was associated with the pellet. From Ho and Varner (unpublished data).*

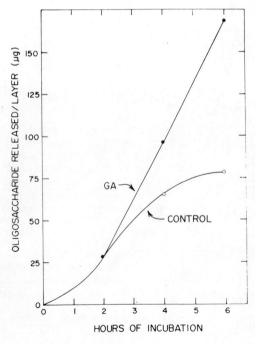

Fig. 22. *Time course of the effect of GA_3 on the release of acid-hydrolyzable oligosaccharide from isolated aleurone layers. From Chrispeels et al., 1973.*

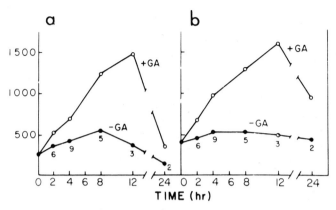

Fig. 23. *The effect of GA on phosphorylcholine-cytidyl transferase activity. Fifty half-seeds were incubated. Aleurone layers were isolated and the enzyme assayed in pellets prepared from cell-free extracts. Each set of points represents an average of a number indicated immediately below each control (data) point of separate experiments. A. 11,000g pellet. B. 44,000g pellet. From Johnson and Kende, 1971.*

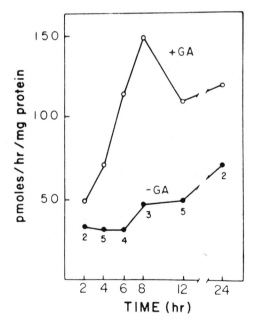

Fig. 24. *The effect of GA on phosphorylcholine-glyceride transferase activity. 50 half-seeds were incubated per treatment as in Fig. 23. The enzyme was prepared as a composite pellet (500g - 44,000g pellet). From Johnson and Kende, 1971.*

REFERENCES

Amrhein, N. (1974). *Planta* **118**, 241–258.
Ashford, A. E. and Jacobsen, J. V. (1974). *Planta* **120**, 81–105.
Bennett, P. A. and Chrispeels, M. J. (1972). *Plant Physiol.* **49**, 445–447.
Ben-Tal, Y. (1974). Ph.D. Thesis. Activation of Phosphorylcholine Glyceride Transferase by Gibberellic Acid in Barley Aleurone Cells. Michigan State Univ.
Ben-Tal, Y. and Varner, J. E. (1974). *Plant Physiol.* **54**, 813–816.
Briggs, D. E. (1963). *J. Inst. Brew.* **69**, 13–19.
Briggs, D. E. (1972). *Planta* **108**, 351–358.
Carlson, P. S. (1972). *Nature* **237**, 39–41.
Chen, R. and Jones, R. L. (1974a). *Planta* **119**, 193–206.
Chen, R. and Jones, R. L. (1974b). *Planta* **119**, 207–220.
Chrispeels, M. J. and Varner, J. E. (1967). *Plant Physiol.* **42**, 398–406.
Chrispeels, M. J., Tenner, A. J. and Johnson, K. D. (1973). *Planta* **113**, 35–46.
Clutterbuck, V. J. and Briggs, D. E. (1974). *Phytochem.* **13**, 45–54.
Collins, G. G., Jenner, C. F. and Paleg, L. G. (1972). *Plant Physiol.* **49**, 404–410.
Eastwood, D., Tavener, R. J. and Laidman, D. L. (1969). *Nature* **221**, 1267.
Firn, R. D. and Kende, H. (1974). *Plant Physiol.* **54**, 911–915.
Gibson, R. A. and Palag, L. G. (1972). *Biochem. J.* **128**, 367–375.
Goodwin, P. B. and Carr, D. J. (1970). *Cytobios.* **7–8**, 165–174.
Ho, D. T.-H. and Varner, J. E. (1974). *Proc. Nat. Acad. Sci. U.S.A.* **71**, 4783–4786.
Ho, D. T.-H. and Varner, J. E. (1976). *Plant Physiol.* In press.
Jacobsen, J. V. (1973). *Plant Physiol.* **51**, 198–202.
Jacobsen, J. V. and Knox, R. B. (1974). *Planta* **115**, 193–206.
Jacobsen, J. V. and Zwar, J. A. (1974). *Proc. Nat. Acad. Sci. U.S.A.* **71**, 3290–3293.
Jacobsen, J. V. and Zwar, J. A. (1975). *Aust. J. Plant Physiol.* In press.
Johnson, K. D. and Kende, H. (1971). *Proc. Nat. Acad. Sci. U.S.A.* **68**, 2674–2677.
Jones, R. L. (1973a). *Plant Physiol.* **52**, 303–308.
Jones, R. L. (1973b). *Ann. Rev. Plant Physiol.* **24**, 571–598.
Jones, R. L. and Armstrong, J. E. (1971). *Plant Physiol* **48**, 137–142.
Keates, R. A. B. (1973). *Nature* **244**, 355–357.
Koehler, D. E. and Varner, J. E. (1973). *Plant Physiol.* **52**, 208–214.
Lin, P. P.-C. (1974). *Adv. Cyclic Nucleotide Res.* **4**, 439–461.
Lin, P. P.-C. and Varner, J. E. (1972). *Biochim. Biophys. Acta* **276**, 454–474.
McNeil, M., Albersheim, P., Taiz, L. and Jones, R. L. (1975). *Plant Physiol.* **55**, 64–68.
Musgrave, A., Kays, S. E. and Kende, H. (1972). *Planta* **102**, 1–10.
Nadeau, R., Rappaport, L. and Stolp, C. F. (1972). *Planta* **107**, 315–324.
Palmer, G. H. (1973). *J. Inst. Brew.* **79**, 513–518.
Varner, J. E., Flint, D. and Mitra, R. (1975). Workshop on Genetic Improvement of Seed Proteins. National Academy of Science/National Research Council. In press.
Varner, J. E. and Mense, R. M. (1972). *Plant Physiol.* **49**, 187–189.
Yomo, H. and Varner, J. E. (1971). *In* "Current Topics in Developmental Biology" (A. A. Moscona and A. Monroy, eds.), Vol. 6, p. 111–114. New York, Academic Press.

Subject Index

A

Abscisic acid
  control of α-amylase synthesis, 177–179
  inhibition of phospholipid synthesis, 179
Actin
  in aggregation deficient mutants, 91–93
  mRNA, 87
  synthesis in Dictyostelium, 86–87
Adenovirus mRNA, 69–70
Aleurone cells, 176–177

B

Barley seedling
  gibberellic acid (GA_3) effect, 173–174
  phospholipid synthesis, 179–188
  protein synthesis, 174

C

Chromatin
  chick oviduct, 125–127
  duck reticulocyte, 11
  high affinity acceptors, 144–146
  initiation sites for,
    E. coli polymerase, 107–124
    oviduct polymerase, 120–124
  in vitro transcription of,
    chick oviduct, 125–131
    mouse globin, 15–26
    ovalbumin, 121–124
  mouse brain, 16
  mouse erythroid, 15–26
  mouse fetal liver, 15–17, 21–24
  nuclease digestion of, 3–7
  RNA initiation sites, 113–118, 129–131
  RS-complex formation, 109–114
  subunit structure, 3–8
Chromatin reconstitution, 7, 15–24

Chromatin structure
  calf thymus, 3–12
  duck reticulocyte, 5–6
  effect of polylysine, 11–12
  globin gene, 11–12
  role of histones, 6–12
Chromatin transcription
  effect of
    estrogen, 107–134
    histones, 127–129
    non-histones, 127–129
  reconstitution, 21–24
Complementary DNA (cDNA), 15–16, 20, 116–118, 123–124

D

Diethylstilbesterol, 109–116
Dexamethasone, 162
Dictyostelium discoideum, 75–78
  HnRNA, 81–83
  life cycle, 75–78
  mRNA, 78–85

E

E. coli RNA polymerase, initiation sites, 118–124
Estradiol
  receptors, 139–140
  vitellogenin synthesis, 153–164
Estrogen
  effect on transcription, 109–112
  receptor, 137–138
    cytoplasmic, 140
    in rat pituitary, 139–140
    in rat uterus, 141–148
    nuclear, 118, 140–141
    response to water excluding forces, 141–143

SUBJECT INDEX

Estrogen-receptor complex
  effect on
    chromatin, 137–139, 144–145
    cytoplasmic exclusion model, 143–144
    genetic expression, 137–138
  nuclear, 141–146
  translocation of, 141–143

G

Gibberellic acid (GA_3)
  effect on
    aleurone cells, 173–174
    α-amylase synthesis, 190–191
    membrane proliferation, 184–188
    phospholipid synthesis, 182–188
    protein synthesis, 173, 177–178
    sucrose release, 181
Globin cDNA, 15–16, 20
Globin cDNA:RNA hybridization, 18–19
Globin gene
  regulation of, 15–23
  transcription, 11
Globin mRNA, 15–16

H

Histones, 3–12
  duck reticulocyte, 7–8
  fractionation, 23–24
  organization of in chromatin subunits, 8–10
HnRNA, 54–61
  adenovirus, 68–69
  Dictyostelium, 81–86
  globin, 55–56
  methylation, 65–66
  poly(A) content, 57–61, 83
  precursor nature of, 66–68
  processing, 85–86
  synthesis, 33–34
hormone receptor complex
  high affinity, 141–146
  nuclear, 146–148
  translocation of, 137–148

M

Methyl groups, in mRNA, 62–70, 80–81
mRNA
  α-amylase, 177
  capping, 80–81, 87
  globin, 55–56
    methylation of, 61–68
    poly(A) addition, 55–59
    post-transcriptional modification, 51–56
  in Dictyostelium, 78–86
    biogenesis of, 81–83
    DNA:RNA hybridization, 83
    poly(A) content, 78–79, 87–90
    preaggregating cells, 93–97
    $5'$-terminus, 80–81
  methylation, 62–70
  post-transcriptional processing, 53–54
  synthesis by steroid induction, 165

N

Non-histone proteins, 15–16, 22–23
Nucleoprotein subunits, 3–6

O

Ovalbumin
  cDNA, 116–118, 123–124
  mRNA, 116–118, 123–124
Oviduct
  gene regulation in, 107–134
  growth and differentiation, 108
  RNA polymerase, 118–124

P

Phospholipid synthesis, 179
Plant hormones
  abscisic acid, 173
  control of protein synthesis, 173–174
  gibberellic acid, 173
Poly(A) (polyadenylic acid)
  in Dictyostelium, 78–79
  synthesis, 57–61, 178–179
Poly(A) RNA, nuclear, 83
Polylysine, effect on transcription, 11–12
Polysomes, in Dictyostelium, 93–97
Protein synthesis
  in Dictyostelium, 86–97
  induction in uterus, 145–146

R

Reconstituted chromatin, 6–12

SUBJECT INDEX

Rifampicin, 109
RNA polymerases
  activity in differentiation, 47–48
  activity in lytic virus infection, 48–49
  α-amanitin sensitivity, 31
  chain initiation, 109–112
  chromatin transcription, 15–16
  complex formation with chromatin, 109–111
  in *B. mori*, 44
  in *E. coli*, initiation sites, 107–118
  in hen oviduct, 118–124
  in mouse plasmacytoma, 36–47
  in Xenopus
    öocyte, 30
    somatic cells, 31
  multiple forms, 29–50
  regulation, 48–49
  subcellular localization, 34
RNA polymerase I, 33–49
RNA polymerase II, 33–49
RNA polymerase III, 39–49
RNA synthesis, chain initiation, 109–116

S

Staphylococcal nuclease
  digestion of chromatin, 3–5
  digestion of nuclei, 4–6
  nuclear limit digest, 5
Steroid acceptor properties, 146–148
  high affinity, 146
  low affinity, 147
Steroid activity
  in vitro, 154, 157–159, 162–164

T

Transcription
  adenovirus DNA, 34–35
  chick oviduct chromatin, 104–134
  and chromatin structure, 3–11
  effect of histones, 127–129
  effect of non-histones, 127–129
  effect of single strand nuclease, 129–131
  effect of single strand antibody, 129–131
  in Dictyostelium, 83–86
  of fetal liver chromatin, 18–19
    globin gene, 11–12
  mouse brain chromatin, 18–19
  reconstituted chromatin, 21–24
  viral genes, 34–35

V

Vitellogenin
  estradiol induction of, 157–161
  mRNA, 164–165

X

Xenopus laevis liver
  conditions for tissue culture, 156–157, 159–161
  differentiation, 165–167
  mRNA, 164–165
  protein synthesis *in vitro*, 156–157
  vitellogenin induction, 157–159, 163–164

A 6
B 7
C 8
D 9
E 0
F 1
G 2
H 3
I 4
J 5